# The Algal Bowl

*Overfertilization of the World's Freshwaters and Estuaries*

DAVID W. SCHINDLER & JOHN R. VALLENTYNE

# The Algal Bowl

*Overfertilization of the World's Freshwaters and Estuaries*

THE UNIVERSITY OF ALBERTA PRESS

Published by

The University of Alberta Press
Ring House 2
Edmonton, Alberta, Canada
T6G 2E1

LIBRARY AND ARCHIVES CANADA
CATALOGUING IN PUBLICATION

Schindler, D. W., 1940–
    The algal bowl : overfertilization of the
world's freshwaters and estuaries / David
W. Schindler, John R. Vallentyne.—Rev. and
expanded ed.

Includes index.
ISBN 978-0-88864-484-8

    1. Eutrophication. 2. Estuarine
eutrophication. 3. Nutrient pollution of water.
4. Freshwater ecology. I. Vallentyne, John R.,
1926–2007 II. Title.

QH96.8.E9S34 2008    577.63'158
C2008-900628-3

The University of Alberta Press is committed
to protecting our natural environment. As part
of our efforts, this book is printed on Enviro
Paper: it contains 100% post-consumer recycled
fibres and is acid- and chlorine-free.

The University of Alberta Press gratefully
acknowledges the support received for its
publishing program from The Canada Council
for the Arts. The University of Alberta Press
also gratefully acknowledges the financial
support of the Government of Canada through
the Book Publishing Industry Development
Program (BPIDP) and from the Alberta
Foundation for the Arts for its publishing activities. We also wish to acknowledge the support
of the Walter and Duncan Gordon Foundation.

*We dedicate this book to our friend and colleague, the late Richard A. Vollenweider, whose seminal work on eutrophication led to many of the advances in the understanding and management of the eutrophication problem that we describe in this book.* —DWS & JRV

*Jack Vallentyne passed away during the publication of this book, and I would like to extend the dedication of this book to the memory of a friend and mentor.* —DWS

# Contents

# Preface

IN THE FIRST EDITION of this book in 1974, J.R. Vallentyne (JRV) predicted that unless something was done to avert our cavalier treatment of lakes, we would find ourselves living in the middle of an Algal Bowl by the year 2000, with degradation of water comparable with the degradation of land that occurred during the great American Dust Bowl of the 1930s. As we shall show, Vallentyne's dire predictions have largely come true, partly for reasons that he predicted, and partly for reasons such as climate warming and rapid expansion of livestock culture, that were not clearly foreseen at the time. Also, the geographical position of the Algal Bowl in North America has shifted westward. The focal centres of the Algal Bowl in the 1960s and 1970s were on areas of dense population and technology in eastern North America and Europe, where sewage wastes rich in phosphorus and nitrogen blended with water, sun, and warmth to provide essential ingredients of the recipe for cultural (man-made) eutrophication. There are two reasons for the shift in position. Firstly, many of the causes of eutrophication that were described in Vallentyne's book have been mitigated in lakes of eastern North America and Europe. Secondly, eutrophication was largely ignored in western North America, and a substantial shift westward of humans, agriculture, and industry has occurred in the last few decades.

In addition to the eutrophication of freshwater, many estuaries and other coastal waters are now experiencing similar symptoms of the problem for many of the same reasons as freshwaters. Luckily, people don't rely on coastal waters for our drinking water, but they are, or in some cases were, important sources of fish and shellfish, so that their plight deserves public concern.

In this book, we (JRV and DWS) have joined forces to bring eutrophication up to date. The book relies heavily on long-term whole-lake experiments at the Experimental Lakes Area in Ontario, which exposed many of the fallacious beliefs about the eutrophication problem that had been based on smaller scale or shorter term experiments. One of these experiments, on Lake 227, is in its 39th year as we finish this book in 2007.

The eutrophication experiments led to many other whole-ecosystem experiments: on the effects of acid rain, manipulation of food chains, the fates of various radionuclides, the effect of reservoir building on greenhouse gas emissions and increased mercury in fishes, and other topics. Long-term data collected as reference material to use for evaluating the effects of the experimental change have proved useful in their own right, allowing us to assess the effects of climate warming, and forest fires in lake catchments. These will be treated in a subsequent book.

### The Dust Bowl and the Algal Bowl

THE DUST BOWL was caused by over-cultivation of marginally productive land in the high plains area west of the Mississippi River during a period of prolonged drought and depression. It was initiated by a reduction of plant cover and the erosion of topsoil by wind following the clearing of prairie land for agriculture, using techniques that had proven successful in the wetter soils of central and northern Europe. "Black blizzards" clouded the sky from view, depositing wind-blown clays and sands over everything in sight, blinding jackrabbits, ruining car engines, and devastating fertile agricultural lands hundreds of kilometres away. In the central part of the Dust Bowl, some barbers refused to shave their customers, for in two strokes a razor could lose its edge from particles of grit embedded in the skin. Matters became progressively worse until May 12, 1934, in what was probably the greatest dust storm in the recorded history of the Earth, over 300 million tons of topsoil were blown eastward across the North American continent, depositing dirt and grit all along the way. By the time it was over, 7.3 million hectares of agricultural land had been degraded, and 250,000 people were forced to leave their homes in the prairies. For more details, see Bonnifield (1979).

Since the "dirty thirties," North America has had even more severe droughts, including the one that is affecting many western U.S. states and Canadian provinces as we write, but for the most part, they have not caused dust bowl conditions. One reason is that farming practices have changed to conserve soil moisture and to keep plant cover on the land for most of the year to prevent soils from blowing away. Another is that, as a result of the Dust Bowl, attempts at dryland cultivation were largely abandoned in semi-arid areas, except where irrigation is possible. Analogous changes in water-use practices could allow us to control the Algal Bowl in much the same way. We will describe some of the needed changes later.

The Dust Bowl arose from mismanagement of land. The Algal Bowl arises from mismanagement of both water and land, the latter because the mismanaged land generally lies in the catchment of a lake or stream. Water that flows off the land, including the chemicals and pathogens that it contains, eventually end up in a body of freshwater. Thus, the mismanagement of land leads directly to the mismanagement of freshwater.

In North America, the Algal Bowl has a potential for disrupting ecosystems, displacing human populations and causing economic hardship over an area that equals or exceeds that of the Dust Bowl. In many more populous parts of the world, this has already happened: potable freshwater and productive fisheries are a thing of the past. We hope that this book may help to avert such a dismal result in North America.

The two of us have worked on the eutrophication problem for a total of over 80 person-years. We have noticed great scientific progress in understanding the causes and the cures for eutrophication. While considerable attention is now paid to controlling point sources of nutrients, non-point sources are still poorly regulated. Canada, and more recently the United States, under the Environmental Protection Agency's Clean Water Act, now have rather aggressive nutrient control regulations. But sadly, many of the decisions that affect eutrophication are still made by municipal and county governments who are unaware of the scientific basis for the regulations. The problem is that much of the excellent science that would help plan development to minimize the effects of nutrients on our waters is instead accumulating dust on shelves in the massive libraries of Ivory Towers around the world. There is a great need for policy makers at all levels of government and average citizens to understand the eutrophication problem, and the role of human activities in causing it. In this book, we have attempted to explain concepts and results in simple language that can be understood by policy makers and intelligent lay audiences, as well as university undergraduates.

# Acknowledgements

SEVERAL PEOPLE HELPED US to complete this book. Margaret Foxcroft made many revisions to our manuscript and obtained permissions for, and final copies of, many of the figures. Brian Parker constructed or revised several figures. Mats Blomquist, Gregg Brunskill, Jim Elser, Mark Graham, Per Jonsson, Greg McCullough, Stephanie Neufeld, John Smol, Daniel Schindler, Brad Stelfox, Michael Sullivan, Lori Volkart, Steve Walker, Howard Webb, and Alex Wolfe gave us permission to use photographs or figures from their papers and manuscripts. Tom Maccagno furnished historical information about the settlement of the Lac la Biche area. Comprehensive reviews of an earlier draft of the book by William Lewis and John Smol helped us eliminate several glaring errors. Selma Losic provided a helpful review as well. A grant from the Walter and Duncan Gordon Foundation made it possible for us to reduce the costs to produce the book.

# 1 | The Algal Bowl

*The biological classification of lakes and the factors*
*involved in human-caused eutrophication*

## Eutrophication

EUTROPHICATION IS THE WORD used by scientists to describe the over-fertilization of lakes with nutrients and the changes that occur as a result. It is derived from the German word *eutrophe*, which is in turn taken from the Greek word *eutrophia*. Ironically, it means "healthy or with adequate nutrition or development," which we shall show is far from our current understanding. In order to understand the eutrophication problem and why it occurs, it is necessary to know a few scientific terms.

Green plants lie at the bottom of the food chain and convert the energy of sunlight into calories of edible food; it is their abundance and rate of production that primarily determine productivity at all higher levels of the food chain. This abundance is related to the abundance of nutrients, in particular phosphorus and nitrogen as we illustrate later in this chapter. In lakes, much of the production is done by algae suspended in the water column, which are collectively known as *phytoplankton*. There are several types of phytoplankton, including diatoms (Figure 1.1), chrysophytes, cryptophytes, dinoflagellates, green algae (chlorophytes), and blue-green algae (now called

Cyanobacteria by scientists; we will use the terms interchangeably here). Three general lake types are defined, in a continuously rising scale of productivity. These are termed *oligotrophic, mesotrophic,* and *eutrophic* lakes.

If they are deep enough, lakes can also stratify vertically, as the result of temperature differences. This is known as *thermal stratification.* Most swimmers have noticed that, if they do a deep surface dive in a lake in summer, there is a point where they quickly pass from a warm upper layer of the lake into a cold deeper layer. The warm upper layer is known as the *epilimnion.* It is the zone where most plant growth occurs, because high sunlight allows photosynthesis to take place. The cold, deeper layer is known as the *hypolimnion.* In summer, it is isolated from the lake's surface by the *thermocline,* the zone of transition. These terms and the reasons why the conditions develop are explained in more detail in Chapter 3. As plants and animals die in the epilimnion, they slowly sink because of the Earth's gravity, ending up in the cold hypolimnion, where they decompose, consuming oxygen.

*Oligotrophic lakes* have a low nutrient supply in relation to the volume of water they contain. As a general rule, they are deep lakes with average depths greater than 15 metres (about 50 feet) and maximum depths greater than 25 metres (about 80 feet). The waters are clear, the result of low plant growth in the epilimnion, so that light can penetrate far enough to allow some photosynthesis to take place in deeper layers of the water column. Typical oligotrophic lakes have high concentrations of dissolved oxygen in the hypolimnion, because the low photosynthesis keeps the rain of algae into the hypolimnion low, so that there is not much decomposition to consume oxygen. In North America, lake trout (*Salvelinus namaycush*), whitefish (*Coregonus* spp.), and other cold-loving species prefer the cold, oxygen-laden deep waters of oligotrophic lakes in summer.

Some of the larger and better known oligotrophic lakes of the world are Great Bear Lake and Great Slave Lake in northwestern Canada, lakes Superior and Huron in the St. Lawrence drainage basin of North America, Crater Lake and Lake Tahoe in the United States, Lake Geneva (Léman) in Europe, and Lake Baikal in Russia. There are also examples from the tropics and the Southern Hemisphere, such as Lake Tanganyika (which is currently threatened with cultural eutrophication); Lake Titicaca in the Andes between Bolivia and Peru; and Lake Taupo, New Zealand.

The most important things to remember about oligotrophic lakes are that they tend to be deep and transparent, with a low density of plant life in surface waters and a high concentration of oxygen in deep water. A Secchi disc,[1] used to measure transparency as described in Chapter 2, lowered into the water can almost always be seen at depths of 3 metres (10 feet) or more in summer. Most naturally oligotrophic lakes have relatively small, naturally vegetated catchments, and low populations of humans and livestock. Some fishes will not tolerate low oxygen or high temperatures.

The world's most unproductive lakes, where nutrients and algal populations are extremely low, are referred to as *ultra-oligotrophic*. Crater Lake, Oregon; Great Bear and eastern Great Slave Lake in Canada; and the central region of Lake Baikal in Russia would fall in the ultra-oligotrophic category. In ultra-oligotrophic lakes, a Secchi disc can often be seem at a depth of 30 metres or more.

*Eutrophic lakes* lie at the other end of the spectrum. They have a high nutrient supply in relation to the volume of water they contain. As a result of this and other factors, dense growths of phytoplankton occur in the epilimnion. In extreme cases, phytoplankton become so dense that they give lakes the appearance and consistency of a thick pea soup. Mats of rooted plants and filamentous attached algae may also carpet the bottom in shallow-water areas, depending on the competition for light between planktonic (open-water) and benthic (bottom-living) plants. The waters of eutrophic lakes typically become very turbid (cloudy) in summer, as a result of plant growth. A Secchi disc disappears from sight at depths of a metre (3 feet) or less. One plant shades another in the struggle for light.

As a general rule, naturally eutrophic lakes are shallow with average depths less than 10 metres (about 33 feet) and maximum depths less than 15 metres (about 50 feet). If a eutrophic lake is deep enough to thermally stratify and develop a thermocline, dissolved oxygen tends to be depleted in the hypolimnion during periods of restricted circulation, because of the decay of sinking organisms. The process of decay also produces carbon dioxide, methane, ammonium, and phosphate, which as we will see are important elements. As a result of low concentrations of dissolved oxygen in the cold hypolimnion, fish that require cold water, such as lake trout, whitefish, and cisco, cannot survive in eutrophic lakes. The dominant fish at southern latitudes tend to be warm-water species that can tolerate living all year in the epilimnion, such as various minnows, northern pike (*Esox lucius*), walleye (*Sander vitreus*), bass (*Micropterus*), and their close relatives. Successional changes also occur in the other organisms inhabiting aquatic ecosystems as productivity increases and deep water oxygen becomes more scarce. This often causes the deep water "dead zones" that the media report for many eutrophic lakes and estuaries. These anoxic zones clearly do not support organisms that require oxygen, including fish. The result is often massive fish kills, usually of species that require cold temperatures and high oxygen, such as whitefish and tullibee. These often float to the surface or accumulate on lakeshores, leading the uninformed observer to conclude that the affected lakes or estuaries are dying. However, the surface waters of eutrophic lakes and estuaries are not dead, but teeming with forms of plant life that we do not value.

In the early 1970s, it was thought that eutrophication couldn't occur in cold lakes, such as those at high latitudes and altitudes. However, that assumption has since been disproven, with several examples documenting that eutrophication of lakes is largely a function of nutrient concentrations. Eutrophic lakes have been shown to occur even in the high Arctic, where lakes never reach temperatures of over a few degrees Celsius and are covered in ice for all but a few weeks each year. Natural sources of nutrients in such areas are almost always low so that, when eutrophication occurs, humans are usually the cause. For example, Meretta Lake near Resolute on Cornwallis Island in the high Arctic was rendered eutrophic by sewage effluent from the human community.

Examples of large eutrophic bodies of water are Lake Winnipeg in central Canada, the western basin of Lake Erie in the St. Lawrence drainage system, Lake Balaton in Hungary, and Lake Victoria in Africa. The most important facts to remember about eutrophic lakes are high nutrient supply; high production at all levels of the food chain; shallow depth; low transparency due to excessive plant growth in surface waters; and, in lakes that stratify, low oxygen in the hypolimnion.

In extreme cases, where lakes are highly eutrophic and undesirable characteristics are extreme, lakes are referred to as *hypereutrophic*. Even in naturally eutrophic lakes, humans have often increased nutrient supplies enough to cause hypereutrophic conditions.

*Mesotrophic lakes* occupy an intermediate position between oligotrophy and eutrophy. They are intermediate in respect to nutrient supply, depth, biological productivity, water clarity, and oxygen depletion in the hypolimnion. Mesotrophy is just a convenient category for lakes that are borderline between oligotrophy and eutrophy.

Many types of human activity increase the nutrient supplies to lakes, causing lakes to increase in productivity. This is known as *cultural eutrophication*. Cultural eutrophication can occur in oligotrophic or mesotrophic lakes, causing them to become more eutrophic. It can also occur in eutrophic lakes, making them hypereutrophic.

Cultural eutrophication is considered to be a problem in well-to-do, technologically developed areas, because clear waters tend to be valued there for aesthetics, recreation, and water supply. In not so well-to-do areas, the situation is often just the reverse. Water is highly valued for food production. It is often deliberately enriched with nutrients to culture maximum yields of fish and edible invertebrates. The tradeoff is that drinking water is often of poor quality, and water is generally not well suited for recreational uses. To understand what has happened to cause cultural eutrophication during the 20th century, it is necessary to outline the biological basis of the trophic (nourishment) classification of lakes.

## A Brief History of Eutrophication

THE GERMAN WETLAND SCIENTIST C.A. Weber coined the term eutrophic in 1907 to refer to the rich wetlands in areas of Europe that received nutrient runoff from surrounding lands. The term was first applied to lakes by Einar Naumann, roughly a decade later. It became widely used by scientists who study lakes to describe the complex sequence of changes in aquatic ecosystems caused by an increased rate of supply of plant nutrients to water. G.E. Hutchinson (1973) provided a detailed history of the use of the term until that time. When this classification of lakes was first developed, it was largely based on the species of benthic (bottom-living) animals present in deepwater sediments. Some species were found to be more sensitive to low concentrations of oxygen than others. As a result, even today some older limnologists refer to lake types in terms of the species of midge (insect) larvae. A *Chironomus plumosus* lake, for example, is eutrophic; a *Sergentia coracina* lake, mesotrophic; and a *Tanytarsus lugens* lake, oligotrophic (Figure 1.2). In later years, once methods were developed for identifying species of algae, quantifying their populations, and measuring plant productivity, emphasis shifted more to these factors to define the trophic status of lakes. Still later, as modern water chemistry developed, limnologists began to use nutrient concentrations directly as a criterion for classifying lakes by trophic status.

FIGURE 1.1: *The frustules of some common diatoms that are preserved in lake sediments. Top right to bottom right: The genera* Aulocoseira, Stephanodiscus, *and* Tabellaria.

Photographs courtesy of Alex Wolfe, Department of Earth and Atmospheric Sciences, University of Alberta.

 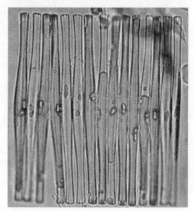

The terms oligotrophic, mesotrophic, eutrophic, and eutrophication were first used in a very descriptive sense based on the appearance of lakes, their oxygen concentrations in deep water, and the species of insect larvae that inhabited deepwater sediments. As the causes of the eutrophication problem and more of its consequences for aquatic organisms were discovered, the meaning of the term has evolved to encompass relationships between a lake and its watershed, energy flow through food webs, and other more dynamic aspects of a lake's condition.

Thus, a single sample of invertebrates dredged from the bottom of a lake at almost any time of the year can give a living history of past conditions of oxygen depletion. In turn, this could be related to the classification of lakes as oligotrophic, mesotrophic, and eutrophic. It appears now, however, that

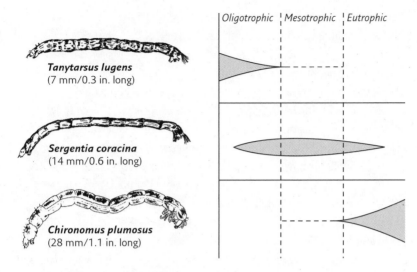

| | Oligotrophic | Mesotrophic | Eutrophic |
| --- | --- | --- | --- |

**Tanytarsus lugens**
(7 mm/0.3 in. long)

**Sergentia coracina**
(14 mm/0.6 in. long)

**Chironomus plumosus**
(28 mm/1.1 in. long)

FIGURE 1.2: *Relative abundance of three species of chironomid (midge) larvae that are representative of oligotrophic, mesotrophic, and eutrophic lakes. The larvae live in bottom sediments, using their gills to obtain oxygen from overlying water.* From Vallentyne (1974). Drawing by Dr. O. Saether.

two factors were confused in old ways of assessing the trophic states of lakes. One of these was nutrient supply, which was primarily determined by events on land. The other was concentration of oxygen in bottom waters, which was largely determined by the form of the lake basin and only secondarily by the rate of nutrient supply.

As limnologists increasingly understood the process of eutrophication, it became clear that there were different species of algae that also typified oligotrophic, mesotrophic, or eutrophic conditions. Diatoms, which build unique cases called *frustules* (Figure 1.1), are particularly useful for deducing a lake's history of eutrophication, because different species are present in oligotrophic, mesotrophic and eutrophic lakes. The frustules of diatoms are formed of pure silica, essentially cases made of glass. The frustules of different diatom species are unique and usually preserved well in most lake sediments, so that the changes in lakes over time can be deduced from diatom fossils in lake muds of known age.

Many of the lakes rendered increasingly eutrophic by human activity have slowly become dominated by Cyanobacteria, commonly known as blue-green algae. Many blue-greens are species that are poor food for animals higher in aquatic food chains and form scums on the lake surfaces.

Some species cause taste and odour problems in drinking water or release toxins that can harm livestock, pets, and people. Although these species leave no hard fossils like diatoms, their pigments are preserved in lake sediments, where their concentrations allow paleolimnologists to deduce changes over time. We will give some examples later in the book.

As criteria for the biological classification of lakes were more fully understood in terms of causal events, it became clear that three primary factors regulated productivity and determined the position of a lake on the scale of oligotrophy to eutrophy. These are now recognized as the rate of nutrient supply, form and depth configurations of the lake basin, and climate (primarily via its effects on light, temperature, and the rate at which water flushes the lake).

In a given climatic zone, natural lakes in areas of nutrient-rich soil were found to be more productive than lakes of similar shape and size in areas of igneous rock drainage. For example, the natural lakes of northern Alberta, which lie in calcareous basins rich in nutrients, are far more productive than analogous natural lakes in northern Ontario, where the basins are of nutrient-poor igneous rock. In regions of comparable climate and geology, tropical lakes are more productive on an annual basis than lakes at higher latitudes, and lowland lakes are generally more productive that lakes at higher altitudes.

Having now discussed the biological classification of lakes, the next step is to consider what happens to natural lakes with the passage of time. On geological time scales, lakes are but a temporary feature of the landscape. Their ultimate fate is to become filled with sediment and eventually be supplanted by wetlands, or even grassed or forested land.

Most of the lakes that now exist were formed recently in terms of Earth history. They were created by the scouring and depositional action of vast continental mountains of glacial ice that blanketed extensive tracts of land in the northern and southern hemispheres as recently as 10,000–15,000 years ago. Wherever a water-impermeable basin was formed by scouring or damming, a lake came into existence. A cursory comparison of maps showing the former distribution of glacial ice and the present distribution of natural lakes suffices to show the relation between the two. For example, Canada, which was largely glaciated, has between two and three million lakes (they are so numerous that no one has actually counted them). The United States, where only the northern states and the mountains were recently glaciated, has fewer lakes, and they are largely concentrated in northern states like Minnesota, Wisconsin, Michigan, and Maine. There are also numerous glacial lakes in the mountains of the western states.

Sediments have been accumulating in these lakes at typically only 1 or 2 millimetres per year. This sounds small; however, accumulated over the 8,000–15,000 years that have elapsed since the recession of the last continental glaciers, many lakes contain 5–15 metres (16–50 feet) of sediments—more in some lakes than in others.

The secrets of the past lie locked in these sediments that have been deposited, particle after particle, over long periods of time. As the sediments have accumulated, pollen and seeds from terrestrial plants, charcoal from forest fires, remains of many insects, crustaceans, and algae are deposited with them, giving a permanent record of the history of a lake and of land-use in its catchment. The information is there for the asking, if one knows what to ask, how to look for the answers, and how to interpret results. *Paleolimnologists* are able to do this from detailed analyses of biological and biochemical remains preserved in sediments. Again, we shall give some examples of this so-called *paleolimnology* later in the book (Chapter 11).

Limnologists once thought that oligotrophic lakes must evolve naturally toward a condition of eutrophy with time, because they slowly fill in with sediment. Logically, this makes some sense, because nutrients from the same catchment area are funneled into less and less water as the lake fills. However, the process would take thousands of years, because of the slow accumulation of sediments in most lakes. Many deep lakes have continuously remained in an oligotrophic state since the time of the last glaciation, suggesting that, under natural conditions, eutrophication takes millennia to occur. Scientific studies that have attempted to document a natural eutrophication process have largely failed, perhaps because other changes to lakes and their catchment are occurring much more rapidly.

In contrast to this slow and perhaps nonexistent process of natural eutrophication, cultural eutrophication has rapidly transformed thousands of lakes in recent decades. It is caused by enrichment of water with nutrients derived from human activities. The principal nutrients involved are compounds of phosphorus and nitrogen. The primary sources are municipal sewage; agricultural and livestock-holding operations; conversion of forested land to grasslands, fields, or cities; and other human sources, including lawn fertilizers and pet excrement. As we shall describe in

Chapter 10, human-caused changes to fish populations, water flows, and climate can also play important roles.

As a result of the enrichment of water with nutrients derived from human activities, numerous waters started to bloom with planktonic algae or growths of aquatic weeds and filamentous attached algae in the 20th century, creating conditions in years or decades that would require thousands of years to come about in the absence of humans. Indeed, many scientists believe that, in the absence of human influence, lakes would not become more eutrophic over time.

As mentioned above, most western countries value clear waters, and the problem of cultural eutrophication that developed during the 20th century was generally considered to be undesirable. As the result of rapid improvements in scientific understanding of the eutrophication problem and some courageous legislative actions, North Americans and Europeans were able to beat back eutrophication's first assault in the mid-20th century. Most European countries now have strong standards to manage nutrients in the catchments of their lakes. Although some lakes are still eutrophic, many of Europe's largest lakes are recovering. The St. Lawrence Great Lakes and many others in eastern North America are also recovering, as a result of aggressive management of phosphate detergents, sewage effluents, and other *point sources* of nutrients.

However, at the turn of the 21st century, eutrophication persists in North America. Phosphate detergents in many areas have diminished in importance as the result of changes to detergent formulations in the 1970s and the adoption of point-source phosphorus control, as we shall describe. Human sewage is still a primary cause of this persistence, but it is reinforced by rapid increases in livestock culture, fertilizer use, land-use change, and disruption to freshwater fish populations and their habitats. As we shall discuss later, there is some evidence that climate warming aggravates the eutrophication of lakes. The Algal Bowl now threatens the region that was once the Dust Bowl.

The sequence of biological and chemical events in this succession is a pattern that repeats itself independently of time and location. The history of Lake Zürich, Switzerland, the first well-documented case of human-caused eutrophication, has since been repeated in thousands of lakes, reservoirs, and estuaries throughout the world. It also occurred near the sites of ancient civilizations millennia ago, such as the Lago di Monterosi, Italy, which paleolimnologists have shown was eutrophied by increased human

FIGURE 1.3: *Cyanobacterial scums washing ashore on the Grand Beach, Lake Winnipeg, August 2006. For colour image see p. 327.* Photograph by Lori Volkart.

activity following the construction of the Via Cassia in 171 B.C. Human attitudes have apparently not changed all that much since the days of the Old
Testament: *"And I brought you into a plentiful country, to eat the fruit thereof
and the goodness thereof; but when ye entered, ye defiled my land, and made
mine heritage an abomination."* (Jeremiah 2:7).

Human-caused eutrophication results in deterioration of water quality
that adversely affects health and economics. Drinking water that was once
clear and pure now tastes "funny" and is turbid (cloudy) during periods
of algal blooms. Filters on municipal and industrial water intake lines
have to be cleaned much more often. Many bathing areas, once clear and
refreshing to swim in, are now coated with algal slimes, and beaches are
occasionally piled high with blue-green algae or other aquatic vegetation
washed in by storms (Figure 1.3). Often, when the algal blooms are caused
by inputs of human sewage or manure from livestock, they are accompanied by high concentrations of coliform bacteria or protozoans such as
*Cryptosporidium* and *Giardia,* which are notorious for causing waterborne
disease. As a result, the cost of water treatment to produce safe drinking
water increases. Increases in toxin-forming algae often occur in summer,

posing even greater problems. In the Third World, typhoid fever and cholera often result from high inputs of sewage to freshwater. Millions of deaths per year result. Unfortunately, the increasing costs of water treatment and the medical costs of increasing waterborne disease are usually ignored when the economics of developments that cause eutrophication are considered.

As a result of increased production of plant and invertebrate food, fish populations often increase in abundance as lakes become eutrophic but with a shift in species composition. Lake trout and whitefish, if present, decline in favour of species of lesser economic value.

Unnoticed by the casual observer, subtle changes in abundance of different aquatic organisms in a lake allow the skilled specialist to extrapolate from the past into the future, to determine both the direction and rate of change. The appearance of certain blue-green algae, such as *Oscillatoria* (now *Planktothrix*) *rubescens,* or the shift from a bottom community of bloodworms (midge larvae) to sludgeworms (aquatic oligochaetes) can be sufficient to characterize the past and suggest changes yet to come.

More commonly, it is not the appearance or disappearance of a species that characterizes the process of eutrophication but complete shifts in species composition of the aquatic ecosystem. Experience, judgement, and intuition are required in interpreting such data. It is not something an untrained person can do any more than a layman can practice medicine.

Naturally productive, eutrophic waters are generally located in regions of naturally productive land. Cases of cultural eutrophication, on the other hand, parallel the distribution and activities of humans. Naturally and culturally eutrophic lakes are often confused if the true causes are not investigated. Also, as we will discuss later for prairie lakes, it is possible to make a naturally eutrophic lake even more eutrophic by adding human activity to the lake and its catchment.

Cultural eutrophication is caused by an increase in the rate of supply of nutrients to an essentially constant volume of water, without any appreciable change in the depth or form of a basin. As a result, cultural eutrophication can often be reversed by eliminating human-caused sources of supply. *Reversed,* however, should not be interpreted to mean anything other than a return toward what was there before the advent of humans. A lake that was eutrophic prior to human settlement cannot be made oligotrophic by removal of human-derived nutrients. At best, it can only return to its former level of eutrophy. If lakes are enriched with enough nutrients, they can be transformed to new stable states, where the reversal of eutrophication can be difficult or even impossible, as we discuss in Chapter 12. An important focus for future lake management must be not to cross the boundaries that cause such undesirable changes in stable states.

In cases where care has been taken to eliminate all human-derived sources of nutrients to nutrient-polluted lakes, signs of recovery toward the pre-settlement state have often appeared within a few years or decades. More often, not all sources of supply are controlled, and results are less than desired. Even when appropriate action is taken, several years may be required before statistical trends of recovery can be distinguished from annual fluctuations caused by weather and other factors. In some cases, recovery is very slow as the result of phosphorus return from sediments. This condition will be discussed in more detail in Chapters 10 and 12.

The most important nutrient elements causing eutrophication are phosphorus and nitrogen. To appreciate their effect in triggering growth of aquatic plants, one need only consider the amounts of phosphorus (P), nitrogen (N), and carbon (C) in typical plant tissues relative to their dry weight and fresh weight. The ratios of these nutrients by dry weight for an average community of algae are approximately 1P:7N:40C. In other words, 1 gram of phosphorus, 7 grams of nitrogen, and 40 grams of carbon would be needed to grow 100 grams of algae, based on their dry weight. The 100 grams dry weight would represent over 500 grams of live algae. This means that, if all other elements are present in excess of physiological needs, phosphorus can theoretically generate 500 times its weight in living algae.

The ratio of algal demand:supply for phosphorus is, on average, higher than for other elements found in plant tissue. There is no atmospheric supply of phosphorus as there is for nitrogen and carbon, the other elements of primary importance to algal growth. Very occasionally, the demand:supply ratio is higher for nitrogen than for phosphorus, as sometimes happens in estuaries. This is discussed in more detail in Chapter 13. Other elements rarely limit algal growth. In situations where phosphorus and nitrogen compounds are in rich supply, as in sewage lagoons, carbon can occasionally become limiting to the rate of algal production. However, even in these circumstances, carbon may not limit total biomass of phytoplankton, because it can be transferred into a lake across the air-water interface from the supply of carbon dioxide in the atmosphere (see Chapters 4 and 9). With regard to other elements required for plant growth, there is no evidence suggestive of roles in any way comparable to those of phosphorus and nitrogen. (However, diatoms—algae with siliceous walls— can be limited by the supply of silicon (Chapter 11), and iron has been found to be limiting in some marine waters.)

Phosphorus and nitrogen, the two elements that most commonly limit plant growth in lakes, are 1000 times more concentrated in sewage effluents than in the waters of lakes unaffected by humans. Therefore, it is not surprising that their addition to natural water has an effect like turning up the volume control on an amplifier, in terms of accelerating the growth of plants. A comparatively small input produces a greatly magnified result. In most cases, algal demand is greatest for phosphorus, so that algae respond most to increased supply of the nutrient. Occasionally, nitrogen is in greater demand, especially in coastal systems (Chapter 13). Other nutrients can modify the species of plants, but the quantity produced is determined by phosphorus and nitrogen.

This amplifying effect of phosphorus and nitrogen compounds has been known to those involved in the pond culture of fish as long as it has to farmers on land. In ponds, phosphate and nitrate fertilizers are added to stimulate growth of algae, the result being a transfer of energy to all higher levels in the food chain, including fish.

In 1970, an average U.S. citizen contributed 1.6 kilograms (3.5 pounds) of phosphorus and 4.5 kilograms (9.9 pounds) of nitrogen to water per year in the form of municipal wastes. More of the phosphorus originated from detergents and industry than from physiological wastes. In addition, whether humans knew it or not, it was estimated that they each contributed about one-tenth of that amount of phosphorus and twice that amount of nitrogen to water through their per capita share of national agricultural and livestock production.

Fortunately, politicians and regulators in North America and Europe heeded the advice of limnologists in the 1970s; since that time, the contribution of nutrients, particularly phosphorus, from cities has decreased, largely the result of elimination of detergents that contained high concentrations of phosphorus and removal of phosphorus from sewage effluent before it is discharged to lakes or rivers.

Unfortunately, the contribution of nutrients from other sources has increased in the past 30 years. Agriculture has become a much more important source, as fertilizer use and livestock culture have increased. There has been increasing emphasis on producing food for sale to parts of the world where high human populations or shortage of freshwater hinder sustainable food production. Runoff of nutrients from paved streets, roads, and mismanaged urban and rural lands also adds nutrients to our fresh-waters. These so-called "non-point sources" of nutrients are much more difficult to control than the makeup of detergents or the point-source effluents from sewage treatment plants.

Quite apart from the use of fertilizers on land, one of the problems arising from agriculture is the "urbanization" of farm animals in feedlots where they are housed and fattened for market. Because the physiological wastes produced by farm animals typically exceed those of the human population in most countries by a ratio of 10:1 or more (for example, it is almost 30-fold more in Alberta, where raising cattle and hogs for export is an important industry), it isn't too hard to understand how eutrophication problems from livestock can arise. With the output of phosphorus from a cow equivalent to that of 11 adult humans, a feedlot with 30,000 head of cattle (not a huge number in the early 21st century) can be equivalent to a moderate-size city in waste output. We will treat this subject in more detail later.

Another factor can be the increasing role of street runoff. It was largely ignored even a few years ago. As urbanization converts more and more land to cities, less and less water is able to penetrate the ground. Instead, rain-fall flows down streets and sidewalks, carrying with it lawn fertilizer, pet excrement, and other sources of nutrients, as well as dozens of toxic chem-icals. Typically, storm drainage is not treated. It is becoming an important source of nutrients in populous areas. It can also be a source of pesticides, herbicides, toxic trace metals, suspended solids, and many other chemicals that can potentially affect freshwater ecosystems. Although results are vari-able, some cities have reduced the problem by increasing street sweeping, limiting use of lawn chemicals, and "poop and scoop" bylaws for pet owners.

One might assume from all that has been said up to now that there is a direct and exact relation between rates at which nitrogen and phosphorus compounds are supplied to lakes and the resulting plant production. Generally this is true but only in a rough, statistical sense. Many factors modify the extent to which nutrients become amplified into plant growth. In addition to rate of nutrient supply, the following can be of considerable importance in determining the extent to which nutrients become expressed in plant growth:

1. the amount of light available to green plants (controlled by light from the sky and suspended clays or dissolved colouring matter in the water);
2. the concentrations and forms of nutrients, which vary with the nature and location of sources of supply (in general, dissolved forms are more available than suspended forms of nutrient);
3. the form and depth of the lake (to be discussed in more detail later);
4. temperature (regulated by geography and climate, with varied effects on different plant species);
5. sedimentation of algae and nutrient-coated clays (varying with turbulence and particle size);
6. removal of nutrients and algae in outflow water (influenced by the "flushing time" of a lake);
7. grazing activities of filter-feeding zooplankton, bottom-living herbivores, and fish, which remove plant food as it is produced (in turn, these are influenced by the populations of predatory fish in a lake);
8. parasitism by bacteria, fungi, and other microorganisms (increased death of plants by disease);
9. regeneration of nutrients from decomposition of plant and animal remains in water and in sediments (reutilization of a former supply); and
10. degree of mixing of lake water by wind (sometimes carrying algae below the photosynthetic zone and also causing upwelling of nutrient-rich bottom water).

It is the interplay of these varied phenomena that attracts limnologists to their task of constructing simplicity and order from what seems like a bewildering complexity.

One final question remains to complete this general account of lakes and eutrophication. Is eutrophication peculiar to lakes? The answer is that it is not. The phenomena associated with eutrophication are usually more easily seen in standing bodies of water such as lakes, reservoirs, and estuaries than in rivers. The algae in rivers are usually attached to the bottom, where they are much less visible than the floating scums that are commonly observed on eutrophic lakes. However, in extreme cases, the attached algae can form mats that are several centimetres thick, causing oxygen depletion when light is too low for photosynthesis to exceed respiration. Under such conditions, oxygen can be depleted overnight, causing suffocation of fish and other organisms. Low oxygen has also been a problem in nutrient-enriched parts of northern rivers, where the river surface is cut off for weeks from atmospheric oxygen by ice cover, and photosynthesis is prevented by snow cover.

Plant production tends to be low in the upper parts of rivers for a number of reasons. On small streams in wooded areas, shading reduces the incident light. Nutrients are low because the area of land drained is small, and they are flushed away before plants have time to utilize them completely; and nutrient concentrate ions are typically low. In the lower stretches of rivers, where nutrient concentrations can be high, suspended silts and clays frequently restrict the penetration of light.

The importance of a second factor of importance, flushing time, can readily be understood by performing a simple kitchen experiment. First, fill an eggcup or cocktail jigger with coffee. Next, place a glass in a large bowl in the kitchen sink. The coffee serves as a pollutant to be introduced into the stream of water flowing from the tap; and the glass, when it is full of water, as a lake. The bowl is only there to collect the overflow, so it can subsequently be measured. Now, start the tap flowing at a medium rate. When the water just fills the glass and begins to overflow, pour the "pollutant" evenly into the stream. The stream will be seen to clear itself instantly; but 8–10 flushings of the glass are required before the liquid within it is restored to its original transparency. The point is that dissolved chemicals in lakes are not flushed out immediately. The resulting delay allows increased time for nutrients to be incorporated into plants and retained within a lake. It can cause a buildup of nutrients from increased inputs due to human activities. Then problems start to arise. The same principle applies when nutrients are added to a lake continuously. For a given rate of nutrient addition, the more slowly water flows through a lake, the higher the nutrient concentration will be.

This completes the general discussion of the classification of lakes as related to eutrophication. In the next chapter, attention is focused on the detailed histories of three lakes unwittingly fertilized by humans and the significance of the water transport system of waste disposal as related to eutrophication.

# 2 | Lakes and Humans

*Human-caused eutrophication caused by increasing populations of humans and their animals and the questionable use of water transport as a means of waste disposal*

ENRICHMENT OF NATURAL WATERS with nutrients induces a variety of biological and chemical changes in water quality. These are usually viewed as detrimental in western countries, because they adversely affect drinking water supply, recreation, and aesthetics. On the other hand, there may be positive effects on fish production, although the species of fish that thrive at high nutrient concentrations are generally not those favoured by sport fishermen.

Four examples illustrate the general sequence of events that takes place when nutrients derived from human culture find their way to lakes. The first pertains to Lake Zürich (Zürichsee), Switzerland, where seemingly trivial events, first recorded in 1896, turned out to be a foreboding of more substantial changes that appeared later. The second concerns four lakes in the vicinity of Madison, Wisconsin, in which massive algal blooms and related events—most unusual to the people living there—appeared soon after Europeans settled in the area. The third refers to the eleventh largest, in surface area, of the world's great lakes, Lake Erie. Our fourth choice is

Lake Winnipeg, which has recently experienced rapid eutrophication and where some special problems impede attempts to recover the lake.

## Lake Zürich

THE AREA AROUND LAKE ZÜRICH has been settled by humans for millennia, from prehistoric lake dwellers through Roman occupation and the Dark and Middle Ages to the present. Looking down on the broad expanse of the long lake basin from the upper part of the old city of Zürich, the colourful panorama, studded with occasional settlements and isolated homes, is impressive. From a distance, it is not evident that the lake has been polluted with nutrients for most of the 20th century nor would one expect the lower part of Lake Zürich to be eutrophic from its rather great mean depth of 50 metres (163 feet) and maximum depth of 143 metres (466 feet). Indeed, the lake was originally very oligotrophic.

So far as has been recorded, nothing unusual happened in Lake Zürich until the 1890s, when two species of planktonic algae appeared in bloom proportions. These were the diatom, *Tabellaria fenestrata,* and the blue-green alga, *Planktothrix rubescens.* This was the signal of later events that, between 1900 and 1970, resulted in massive growths of diatoms and blue-green algae; severalfold increases in concentrations of chloride ions and dissolved organic matter in the water; decreased water transparency in summer; precipitation of calcium carbonate (marl) as a result of increased photosynthesis; growths of filamentous algae, such as *Cladophora* and *Ulothrix,* on bottom areas near the shore; oxygen depletion in the bottom water; disappearance of trout and whitefish populations; and a concomitant rise in the abundance of perch and members of the minnow family.

Dr. L. Minder, one of the early Swiss limnologists, documented many chemical changes that took place. While he was engaged in these studies, another Swiss scientist, Dr. Fritz Nipkow—pharmacist by profession and limnologist by desire—was examining cores of sediment taken from the deepest part of the lake. Nipkow discovered that the sediments contained delicate, well-preserved layers representing seasonal successions of events year after year from the time of the first algal blooms. Pollen grains and diatom frustules associated with early spring were overlain by late spring and summer forms and these, in turn, by species characteristic of late summer and autumn. The same pattern was repeated layer after layer. Because of this fortunate circumstance, it was possible to answer a number of questions pertaining to the lake that would have been difficult to answer otherwise. One of these concerned the abundance of blue-green algae.

Although high abundances of blue-green algae (or algal "blooms" as they are now commonly called) were known to have occurred in Lake Zürich with increasing frequency since the 1890s, few remains occurred in the sediment. Unlike diatoms, which have strong siliceous cell walls, Cyanobacteria tend not to preserve very well. However, a Swiss chemist, Hans Züllig reported finding a pigment peculiar to Cyanobacteria in the sediment in 1959. The name of the pigment was myxoxanthophyll. By determining concentrations of myxoxanthophyll at different levels in the sediment, and with the help of Nipkow's bands, Züllig was able to trace the incidence of Cyanobacteria blooms back to the 1890s.

Studies undertaken after World War II by Dr. E.A. Thomas and other Swiss scientists showed a buildup of the growth-limiting nutrients nitrogen and phosphorus. The nutrient increases occurred in waters below the photosynthetic zone and were in the form of phosphates ($PO_4$) and nitrates ($NO_3$), forms of the elements that are ideal for utilization by algae and other plants. Paralleling this buildup, the incidence and duration of algal "blooms" (as the dense growths of algae have come to be called) began to increase with time. When algae reduced the concentrations of critical nutrients in surface waters to almost undetectable levels in summer, blooms persisted because of episodic supply of nutrients from non-point sources, because of recycling of nutrients from deeper water, and because the species that composed the blooms were too large to be readily eaten by grazing zooplankton.

While all of these changes were taking place in the lower, densely settled part of the lake, the shallow upper basin, separated by a constriction of the lake, was not appreciably affected by humans. It remained much as it had always been, forming a control for the experiment humans had unwittingly performed in the lower lake.

To combat these changes, a drainage diversion was constructed around part of the shore, carrying nutrient-rich wastes to the Limmat River, which was the outlet of the lake and transported the nutrients to the Rhine River and then to the North Sea. In the mid-1960s after a long controversy on the best means of controlling eutrophication, a program of phosphorus removal from municipal wastes was initiated in an attempt to reduce the problems associated with increased plant growth.

By the early 1980s, it was clear that Lake Zürich was recovering from its long history of enrichment with nutrients from municipal and agricultural wastes. Total phosphorus concentrations decreased from over 100 micrograms per litre to less than 20 micrograms per litre in 1999, as the result of elimination of 99% of the input of phosphorus from human sources. An even more striking recovery was observed in Lake Lucerne, another huge oligotrophic mountain lake in Switzerland. As in Lake Zürich, increasing populations of humans and livestock rendered the lake eutrophic in the early half of the 20th century, with highest algal blooms in the mid-1970s. As a result of decreasing phosphorus input from 103 to 14 tonnes per year, the lake became mesotrophic, then oligotrophic again over a period of only 15 years. The recoveries of these large lakes celebrate the successes of 20th-century limnologists and aquatic engineers.

## The Madison Lakes

THE NEXT CASE concerns four lakes in the vicinity of Madison, Wisconsin, USA. Their depths, areas, and volumes are given in Table 2.1. An outline map of the lakes is given in Figure 2.1.

TABLE 2.1: *The depths, areas, and volumes of the Madison Lakes.*

| Lake | Depth (m) Maximum | Mean | Area (km²) | Volume (m³) |
|------|---------|------|-----------|-------------|
| Mendota | 25.6 | 12.1 | 39.4 | 478,000,000 |
| Monona | 22.5 | 8.4 | 14.1 | 119,000,000 |
| Waubesa | 11.1 | 4.9 | 8.2 | 40,000,000 |
| Kegonsa | 9.6 | 4.6 | 12.7 | 59,000,000 |

The history of the Madison lakes is a typical example of what has happened to lakes in the United States and southern Canada, and we shall describe it in some detail, because many of the same mistakes that caused the eutrophication of the lakes are still being made today. It is also an example of how complex the process of recovery can be, once many stakeholders have vested interests in a lake or its catchment.

The Madison lakes drain several small streams, joining as a chain to become the Yahara River, which drains to the Rock River, a tributary to the Mississippi. Originally, the area around the lakes was forested, with some oak savannahs, prairies, and wetlands. Geologically, the area is calcareous,

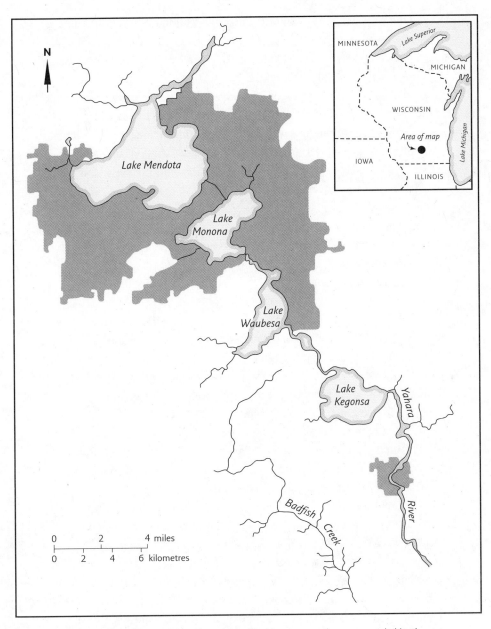

FIGURE 2.1: *The Madison lakes. The area outlined in gray was the area occupied by the city of Madison in 1970.* Reprinted from Vallentyne (1974). Map by Wendy Johnson.

and the lakes have hard waters i.e., high concentrations of calcium, magnesium, and bicarbonate.

John Wakefield was one of the first Europeans to view the lakes, passing in 1832 as he participated as a soldier in the Black Hawk war against the Sauk Indians. In his 1834 account of the war, he described them as "... the most beautiful bodies of water I ever saw." By the late 1840s, land clearing for agriculture was well under way in the catchments of the lakes. Within four decades, farms had replaced native vegetation, and Madison was a prosperous town. By 1880, newspaper accounts regularly reported algal blooms and fish kills in the lakes. Land-use changes and raw sewage discharged into Lake Mendota by 10,000 inhabitants of Madison at that time were the probable causes. Dams were also installed to control water levels, wetlands were drained, and wells were drilled to tap groundwater in the catchments. The level of the largest lake, Lake Mendota, was raised by 1.5 metres by a dam built in 1847, submerging many of the nearshore wetlands.

E.A. Birge arrived in the 1880s to establish a limnology program at the University of Wisconsin, one of the first in the United States. He and his colleagues began to document changes to the lakes. However, it would be a century before the intimate links between changes in the lakes and the modification of the surrounding landscape were well understood.

The first attempt at sewage treatment in Madison was initiated in the latter part of the 19th century. A new treatment plant to handle the growing population was later put into operation in 1914, with the treated effluent discharged into the Yahara River just above Lake Monona. The sewage, as usual, was treated for the removal of fast-settling solids and organic carbon, but not for removal of phosphorus or nitrogen. Within 2 years of installation of the treatment plant, it was obvious that something quite unexpected (by the local population) was taking place. Blooms of blue-green algae, particularly *Anabaena* and *Aphanizomenon,* were appearing in the lake in frequency and abundance never seen before. By the early 1920s, thick mats of floating blue-green algae, piled up by onshore winds, were so offensive that even on hot and humid nights persons living near the lake bolted their windows, preferring a stifling enclosure to the putrid breeze.

From 1912 to 1953 copper sulfate, an algal poison, was applied to the surface waters of Lake Monona in an attempt to get rid of the algae. Between 1926 and 1936, 27,000–45,000 kilograms (60,000–100,000 pounds) of copper sulfate were applied annually to reduce algal growth. These additions now lie buried in copper-rich layers of sediment—a useful reminder to future scientists of the futility of trying to solve human-caused eutrophication problems without cutting off the supply of nutrient-rich wastes.

In 1928, the city of Madison completed the first stage of a new sewage treatment plant. Effluent from this plant bypassed Lake Monona, entering the Yahara River between Lake Monona and Lake Waubesa. In other words, nothing had really been done other than to shift the effluent to the next lake downstream. Eutrophication problems immediately started to appear in lakes Waubesa and Kegonsa. Copper sulfate was, in turn, applied to these lakes to reduce algal growth. Lake Monona improved, but not as rapidly as expected. This was because of continued, although lower, input of sewage from developing upstream communities, runoff from agricultural areas, storm sewers, and industrial effluents that still entered Lake Monona and Lake Mendota.

Later work indicated that the Yahara River below the four lakes was also affected. In 1954, a dense bloom of the blue-green alga, *Aphanizomenon flos-aquae,* caused a complete fish kill below the point of discharge of the sewage effluent. As a result of intense biological consumption of dissolved oxygen at night, at a rate faster than it could be replenished from air, the fish died of asphyxiation. Only a few carp (*Cyprinus carpio*) survived in a small pocket above the point of discharge.

The history of biological events in the Madison lakes in the first half of the 20th century is not as completely recorded as for the Lake of Zürich, but enough is known to say with assurance that the causes and responses were similar in both cases. In 1944, Claire N. Sawyer and James B. Lackey showed that the causal factors of eutrophication in the Madison lakes were supplies of phosphorus and nitrogen compounds in the incoming waters. In particular, phosphorus in the lakes increased rapidly in the 1940s. By the 1960s, much public concern focused on Lake Mendota. Sewage from the upstream communities was diverted from the lake in 1971, decreasing phosphorus inputs by about 30%. However, non-point sources of nutrient were ignored. The increasing use of synthetic fertilizers following World War II to support increasing culture of corn caused increases in nutrient runoff. Soils of the Lake Mendota catchment have been increasing in phosphorus at the rate of 575,000 kilograms per year since the 1970s. The reason is that more phosphorus is imported to the basin as fertilizer and animal feed than is exported as produce. As a result, the lake showed little improvement, despite the diversion of sewage. Internal recycling of phosphorus from sediments to overlying water also became very efficient during summer stratification as the decay of plant matter consumed oxygen. In some years, recycled phosphorus actually exceeded new sources from the catchment. Later, we shall describe how oxygen and sediment return are related and show that this "internal loading" usually decreases slowly if external loads are reduced.

The biological community of the lakes was also greatly modified. Invasion by the rooted aquatic macrophyte known as Eurasian watermilfoil (*Myriophyllum spicatum*) displaced native species of aquatic plants. Introduction of common carp intensified eutrophication by mixing of bottom sediments. The increased turbidity further reduced beds of native macrophytes. Freshwater drum (*Aplodinotus grunniens*) and yellow bass (*Morone mississippiensis*) were also introduced. The Eurasian cladoceran *Eubosmina coregoni* appeared. Most of the native species of small littoral fishes disappeared as the diverse beds of native macrophytes disappeared. Waterfowl that fed on native macrophytes declined in abundance. Fingernail clams disappeared. Heavy fishing pressure all but extirpated piscivorous (fish-eating) fish species such as pike (*Esox lucius*) and walleye (*Sander vitreus*), a very common problem as human populations increase. As we shall discuss in Chapter 9, removing piscivorous predators can initiate a *trophic cascade*, which can amplify the production of algae by nutrients.

Even in the mid-20th century, few scientists connected the damage to fisheries with the eutrophication problem, but observations by Dr. J. Hrbáček in Czech fishponds and Dr. John Brooks and Dr. Stan Dodson in eastern North American lakes suggested otherwise. Dr. S.R. Carpenter and Dr. J. Kitchell first formalized the pattern, where lakes at a given nutrient concentration with an even number of trophic levels would have lower algal abundance than lakes with an odd number. They and their colleagues conducted whole-lake experiments in northern Wisconsin, which showed that "biomanipulation" could improve water quality, a subject that we will explore in more detail later in the book. These experiments suggested that restoring the piscivorous species to Lake Mendota would reduce populations of zooplanktivorous minnows and other small fish, allowing grazing zooplankton to increase. This would result in more efficient grazing of the algal community of the lake.

During the period 1987–1999, 2.7 million walleye (*Sander vitreus*) finger-lings, 170,000 northern pike (*Esox lucius*) fingerlings, and 100 million walleye and northern pike fry were added to Lake Mendota. The combined biomass of the two species ranged from 4 to 6 kilograms per hectare, about average for northern lakes.

However, word soon spread in the angling community. Fishing pressure on walleye populations increased by sixfold in 1987–1989 and remained high through the 1990s, which diminished the impact of the biomanipulation.

Even so, water clarity improved during the early years of the biomanipulation. *Daphnia pulicaria*, a large, efficient grazer, replaced the smaller, less effective *Daphnia galeata mendotae* as the dominant grazer. However, it is not clear whether this change occurred because of the biomanipulation or because of a 1988 summerkill of cisco (*Coregonus artedi*), the predominant zooplanktivorous species in the lake.

As the city of Madison and other urban and suburban developments have increased in the catchment of the Madison lakes, concern has arisen for nutrients carried in by the storm drains that catch runoff from city streets. Although the urban and suburban areas are still smaller parts of the catchment than agriculture, their yields per unit area can be as high or higher. As a result, there have been recent measures to reduce inputs of nutrients to the lakes from storm drains.

In summary, although the water quality of the Madison lakes is not as bad as it was in the early years of the 20th century, algal blooms, fish kills, and other symptoms of eutrophication continue to appear. Recent studies of Lake Mendota indicate that improving the water quality could be worth $50 million in lost ecosystem services, maintaining public interest in continuing restoration efforts.

Efforts to restore Lake Mendota have returned to focus on non-point nutrient sources. They are slowed by having to deal with multiple bureaucracies, and a large number of individual landowners. Steve Carpenter describes attempts to control eutrophication as an "ongoing experiment." However, eutrophication has gone on so long that soils in the catchment of the lake are saturated. Carpenter (2005) estimates that even if sources of fertilizer and manure to the catchment were removed, soils would continue to leak elevated concentrations of phosphorus for several decades.

## Lake Erie

AS INTERESTING as these histories of lakes in Switzerland and Wisconsin may be, they are dwarfed by the geographic immensity of change in Lake Erie, fourth largest in surface area of the chain of five St. Lawrence Great Lakes—a body of water 26,000 square kilometres (almost 10,000 square miles) in area. It is ironic that Lake Erie, the last of the St. Lawrence Great Lakes to be penetrated by Europeans, was the first to fall prey to their machinations.[1]

When Louis Jolliet and his successor, Sieur de La Salle, viewed the shore of Lake Erie from their heavily laden canoes in 1669, they found the water as rich in the variety and size of fish as was the lush forest canopy on the shore with woodcock, quail, wild turkey, and deer. When Sieur de Cadillac penetrated to the present site of Detroit in 1698, he had to part wild geese and swans with his canoes in order to land his party on shore. Now extinct, the passenger pigeon seasonally darkened the sky over Lake Erie by sheer immensity of numbers.

As 18th-century exploration and fur trade gave way to settlement and industrialization, environmental damage began to spread from harbours and bays into the main body of the lake. Events that set this into motion started with the formation of coal deposits in Pennsylvania, Kentucky, and Ohio 300 million years ago and with the formation of enormous deposits of iron ore along the southern shore of what is now Lake Superior over 1 billion years ago.

Following the discovery of these deposits by technological humans, Lake Erie became the natural centre of the U.S. steel industry. The reason was the ease of water transportation and the 4:1 ratio of coal to iron ore then used in the manufacture of steel. The economics of the 4:1 ratio demanded the centre of the steel industry be on Lake Erie rather than Lake Superior. As the gross tonnage of iron ore passing through the locks at Sault Ste. Marie increased from a few thousand tonnes in 1855 to hundreds of millions of tonnes in 1973, so did the pollution of Lake Erie from industry and human population.[2]

Adequate support for comprehensive limnological studies on the Great Lakes was not made available by the governments of either the United States or Canada until the 1960s. Prior to that time, Paul R. Burkholder, Charles C. Davis, and David C. Chandler, to name but three of the early limnologists, performed their work in the 1930s and 1940s largely by "rowboat" limnology.

When adequate financial support did come in the 1960s, the hour was late. Such profound changes had taken place that it was difficult to determine what the lake had been originally been like. The available scientific information, collected with inadequate facilities by a handful of dedicated scientists, was extremely limited.

Unprecedented changes had taken place in the fish populations. Between 1930 and 1965, the numbers of cisco (*Coregonus artedi*), sauger (*Sander canadensis*), lake whitefish (*Coregonus cluepeaformis*), yellow walleye, and blue walleye in the commercial harvest decreased in succession. By 1970, it could be said authoritatively that blue walleye, once a major species, was extirpated from the lake. Yellow perch (*Perca flavescens*), rainbow smelt (*Osmerus mordax*), and white bass (*Morone chrysops*), on the other hand, were increasing.

It is virtually impossible now to reconstruct the causes of the changes in fish populations because of lack of fishery and limnological information. Overfishing, oxygen depletion in bottom water, changes in abundance of food organisms, introduced species, toxic pollutants, and ruination of spawning beds were probably all involved. The only clearly identifiable cause was humans.

From old records of algal populations at a municipal water intake plant at Cleveland, Ohio, Dr. Charles C. Davis, then at Case Western Reserve University in Cleveland, Ohio, discovered algae had been increasing rapidly since 1930. Typical results in 1927–1930 were 100,000–200,000 cells per litre. By 1946–1948, comparable values had risen to 1,200,000 cells per litre; and by 1960–1964 had increased to 1,300,000–2,400,000 cells per litre. Diatoms and Cyanobacteria were recorded in what seemed to be unprecedented numbers at numerous points in the lake.

By the mid-1950s, the shallow western basin of Lake Erie, loaded with wastes from Toledo, Detroit, and other major cities on the U.S. side, had become dangerously susceptible to any change that might tip the scales of nature in an unpredictable way. The first event was a period of unusually warm and quiet weather from September 1 to 5, 1953. As a result of water stagnation and consequent oxygen depletion caused by the decay of algae at the mud surface, practically the entire population of a mayfly nymph, *Hexagenia*, was wiped out in the western basin. This mass mortality of an important item in the diet of many Lake Erie fish was caused by the biological richness of the water and sediments—in part due to nature, in part due to humans.

After a brief attempt at recovery, populations of *Hexagenia* disappeared in the western basin by 1960 and were drastically reduced throughout the rest of the lake. To those living around the shores of Lake Erie, it was a mixed blessing, because they were no longer pestered by swarms of mayfly adults around light bulbs and screens at night. However, the fish were not so pleased; nor were the fishermen whose livelihood depended on fish.

In the early 1960s, Dr. Alfred Beeton of the U.S. Bureau of Commercial Fisheries (later at the University of Wisconsin, Milwaukee) showed that the chemistry of Lake Erie had changed dramatically since the early part of the century. Comparing concentrations of different chemicals reported by various investigators, he showed that concentrations of sodium, chloride, and sulfate in Lake Erie had more than doubled since 1910. Furthermore, these changes were passed downstream to Lake Ontario. In contrast, concentrations of chemicals in Lake Superior (uppermost, largest, and least densely settled of the St. Lawrence Great Lakes) had not varied during the same period.

In 1961, Dr. John Carr and Dr. James Hiltunen of the U.S. Bureau of Commercial Fisheries compared the kinds and abundance of bottom invertebrates in the western end of Lake Erie to those reported for 1930. They found the polluted area of the western basin had increased fourfold from 263 square kilometres (101 square miles) in 1930 to 1020 square kilometres (388 square miles) in 1961. In the 1960s, U.S. Coast Guard authorities advised international shippers not to draw nearshore water from the western end of Lake Erie for drinking purposes, because of possible danger to health.

Biological and chemical changes in Lake Erie and Lake Ontario were more pronounced than in the other Great Lakes. The reason for this in Lake Erie is its relatively shallow depth, small volume (the smallest of all the St. Lawrence Great Lakes), and the 12,000,000 people inhabiting the drainage basin in 1966.

Many people wrongly referred to Lake Erie in the 1970s as a dead or dying lake. This mistake was probably the result of the fish kills observed during summer, when decomposition robbed the deep waters of the lake of oxygen. The real problem was that the western and central basins of the lake were teeming with life (Figure 2.2), in the form of unwanted microorganisms like blue-green algae and coarse fish. Dr. Andrew L. Hamilton and Dr. William W. Warwick of the Freshwater Institute in Winnipeg showed from the analysis of sediment cores that the western basin of Lake Erie was

FIGURE 2.2: *Lake Erie from the air in 1969, showing a massive bloom of Cyanobacteria or blue-green algae. For colour image see p. 327.* Photograph by DWS.

in a eutrophic state prior to the appearance of technological humans on the shore, but they also showed that, since the advent of humans, the degree of eutrophy had markedly increased. The "overload" was all due to humans.

Mixed with more than occasional buckets of oil and grease (1000 barrels a day in the Detroit River alone in 1967), organochlorine insecticides, PCBs, detergent phosphates, mercury, and the wastes of municipalities, industries, and ships, the western two-thirds of the lake was in the process of becoming an enormous sewage lagoon. Open waters in the eastern end of Lake Erie, because of their greater volume and depth, retained comparatively clear relative to the rest of the lake. When the Cuyahoga River at Cleveland caught fire in 1969 from oil on its surface, it was painfully obvious to humans that they had severely damaged the lake.

Like the great mountain lakes of Switzerland, Lake Erie also recovered as phosphorus inputs were decreased. Phosphorus and chlorophyll *a* in the once-eutrophic western basin of the lake decreased from 41 micrograms per litre and 14 micrograms per litre, respectively, in the mid-1970s to <20 micrograms per litre and 5.6 micrograms per litre in the 1990s (Charleton and Milne 2004). Inputs of many other pollutants were reduced, making the lake less toxic. Lake Erie became clear again.

Late in the 20th century, Lake Erie again began showing other signs of stress. Another consequence of human activities has been the invasion of the lake by dozens of alien species of fish and invertebrates. For invertebrates, this is largely the result of ships arriving from other continents dumping their ballast water in the Great Lakes. Zebra mussels (*Dreissena polymorpha*), an alien mollusc species from Eurasia, and other alien species of molluscs have invaded the lake, replacing native mussels. These mussels can filter large volumes of lake water each day. In the process, they take nutrients from the water column and excrete them along the lake bottom in littoral areas where the mussels live. The result has been a renewed proliferation of *Cladophora* a genus of filamentous green algae that washes ashore on strong winds to rot in windrows on the beaches. This transfer of nutrients from the water column to the sediment surface, termed the *near-shore benthic shunt,* has been described by Dr. Bob Hecky (2004).

## Lake Winnipeg

LAKE WINNIPEG is nearly the same size as Lake Erie. However, it drains a huge but largely semi-arid catchment that extends from the Rocky Mountains of the west to well into northern Ontario. Human communities on the lakeshore are few, mostly small fishing communities or aboriginal communities (First Nations to Canadians). The Red River enters the lake from its south end, draining a rich agricultural area of Minnesota, North Dakota, and southern Manitoba. In addition, it receives sewage and street runoff from Winnipeg (which has over 600,000 people) and several sizeable American towns and cities. The Winnipeg River, which runs from culturally eutrophied Lake of the Woods to enter Lake Winnipeg's southeastern corner, is another important nutrient source.

Lake Winnipeg had many names, conferred by the various First Nations that lived on its shores to utilize its rich fishery. It was Mistehay Sakahegan to the Cree, who arrived from the east about 2000 years ago, and Gitchi gumee to the Ojibwa, who arrived in the 1700s to participate in the fur trade. Both names simply mean "Great Lake." To the Assiniboine of the western plains, it was Men-ne-wakan, or "Mysterious Water." Earlier fur traders called the lake by a variety of descriptive names. The present name, conferred by Europeans in the late 18th century, is thought to originate from the Cree word *Winnipak* for the muddy colour of the lake's water.

Humans have lived in the Lake Winnipeg basin for at least 8000 years, only a few thousand years after the basin was a part of the immense glacial

Lake Agassiz. Henry Kelsey was probably the first European to see the lake, crossing it on a 1690 voyage from York Factory on Hudson's Bay to the current site of The Pas, upstream on the Saskatchewan River. With its immense drainage, the lake became a crossroads for fur trade activity, with canoes approaching from Hudson's Bay via the Hayes and Echimamish rivers, from the St. Lawrence Great Lakes via portages into the Rainy and Winnipeg rivers, from the south via the Red River, and from the west via the Saskatchewan and Assiniboine rivers. The settlement Lower Fort Garry, a few kilometres south of the lake on the Red River, was an important early Hudson's Bay Company post. It was replaced by Upper Fort Garry, several kilometres upstream at the junction of the Assiniboine and Red rivers, which was to evolve into the modern city of Winnipeg.

The lake became a site for settlement by Icelanders in 1875, when Sigtryggur Jonasson brought 285 settlers to the lakeshore near Gimli. In 1876, 1200 more Icelandic settlers followed. The settlers formed several small communities on the lake's shore to exploit its fishery. Much of the lake's watershed remains sparsely populated. Recent reports indicate that 23,000 people live in 30 communities on the lake's shore. Dispersed through Lake Winnipeg's large catchment are 5.5 million Canadians and 1.1 million Americans.

Until the 1970s, most of the travel among communities on northern Lake Winnipeg was by boat. A lucrative fishery for walleye, lake whitefish, and other species was the main livelihood of residents. Some of the more remote communities on the eastern shore are still not connected to "outside" by roads, but a network of new logging roads is changing that rapidly.

The lake was first studied by A. Bajkov in 1929. He described most of the fishes and planktonic invertebrates of the lake. In 1969, a complete survey of the lake was organized by Dr. Gregg Brunskill of the Freshwater Institute in Winnipeg. Both JRV and DWS were part of Gregg's team (Figure 2.3), which did several cruises using the *Bradbury,* an older ship that was then used to tend navigational buoys on the lake. Captain Chris Thorsteinsen was in command.

The study found that the southern basin of the lake, which received the nutrient-rich inflow from the Red River, was quite eutrophic. However, some of the worst symptoms of eutrophication, associated with the depletion of oxygen in bottom water, did not occur because the lake was so windswept that the thermoclines formed only rarely, and the water column

FIGURE 2.3: *Authors DWS (top) and JRV inspect a newly constructed phytoplankton production incubator aboard the* Bradbury *before the first cruise on Lake Winnipeg in June 1969. The* Bradbury *is now in the marine museum at Selkirk, Manitoba, where it can be toured by the public.* Photograph courtesy of G. Brunskill.

is usually well mixed. The north basin of the lake was also quite productive, with considerable production of Cyanobacteria in late summer.

No comprehensive work on the lake was done again until the 1990s. By that time, the *Bradbury* had been retired to the maritime museum in Selkirk, Manitoba. In 1991, a consortium of scientists from the Winnipeg area, funded largely by Manitoba Hydro, re-surveyed the lake, using another elderly ship, the *Namao*. In the first summer of the study, the lake appeared to have changed little since 1969. However, it changed rapidly in the 1990s. Fertilizer use increased in the basins of the Red and Winnipeg rivers, and many intensive livestock operations were begun in the southern part of the lake's catchment. Cities continued to discharge effluent into the lake via the Red River. A 1997 flood in the Red River basin carried vast amounts of nutrients, pesticides, and silt into the lake. Meanwhile, drought in the western prairies and dam construction reduced water inputs from the Saskatchewan River. The Saskatchewan River is relatively low in nutrients, and in earlier years, it diluted the high-nutrient waters from the southern inflows. The outlet of Lake Winnipeg was partially dammed to control flows to hydroelectric installations, increasing retention of nutrients in the lake. As a result, phosphorus, nitrogen, algae, and zooplankton all increased greatly. In a very few years, Lake Winnipeg became as eutrophic as Lake Erie was in the early 1970s. In 2003, satellite photos revealed a huge mat of blue-green algae, 6000 square kilometres in area, that covered the lake for three months (Figure 2.4). Scientists of the Lake Winnipeg Consortium continue to try to unravel the complexity of changes that have caused the lake to become eutrophic so rapidly. As we write (2007), no decisions have been made about controlling the sources of nutrients in the catchments of the Red and Winnipeg rivers. In addition, several alien species of fishes and invertebrates have begun to appear in the lake. Recently, a pipe was constructed to carry saline water from Devils Lake, North Dakota, to the Sheyenne River, a tributary to the Red River. It would then flow to Lake Winnipeg, where many believe that it provides still another threat to the lake. The flow of water is currently held up by a technicality: the concentrations of sulfate exceed the water quality guidelines of the state of North Dakota. Lake Winnipeg is desperately in need of attention. Ignorance is no longer any excuse for allowing lakes to deteriorate from eutrophication.

A recent (2006) report from the Lake Winnipeg Stewardship Board describes in some detail the problems of the lake. Over 200 small wastewater treatment plants and 10 larger ones discharge sewage to the lake. Growing livestock culture and increasing fertilizer use are other problems.

In summary, there are many case histories of eutrophication. Some of them are success stories resulting from reducing phosphorus inputs to lakes in the United States, Canada, and Europe. For example, see papers in the October 2005 issue of the journal *Freshwater Biology*, available at most

FIGURE 2.4: *A satellite photograph of Lake Winnipeg (right) in September 2005. Colours ranging from browns through yellows to greens indicate increasing vegetation density or vigour. In the case of Lake Winnipeg, greens are due to floating blue-green algae (i.e., surface blooms) almost indistinguishable from the green of the vegetation in the forest on either side of the lake. White indicates clouds. The surface blooms have reoccured for the last several years. Lakes Winnipegosis and Manitoba are to the left of the lake. The red-brown areas in the south basin of lakes Winnipeg and Manitoba are caused by silt suspended by wind, the result of its shallow depth. Similar observations have been made in all recent summers. For colour image see p. 328.*

Normalized difference vegetation index map supplied by Greg McCullough, University of Manitoba. Created from National Aeronautics and Space Administration MODIS (Moderate Resolution Imaging Spectroradiometer) bands 1 and 2 data.

university libraries. We know most of what we must do to prevent eutrophication and to recover culturally eutrophied lakes. What is often missing
is the courage of policy makers to apply the measures that we know well.

## Nutrients and Waste Treatment

To probe more deeply into the underlying causes behind these and
similar events taking place in tens of thousands of lakes, reservoirs, and
estuaries throughout the world, there is no better path to follow than the
history of methods used by human populations for disposal of fecal and
urinary waste.

The earliest method adopted for disposal of physiological waste was
by addition to the soil, a time-honored practice still used in many parts of
the world today, including the United States and Canada. It was this same
practice Moses advised the Israelites to follow when he said, (Deuteronomy
23:13) *"And thou shalt have a paddle upon thy weapon; and it shall be, when
thou wilt ease thyself abroad, thou shalt dig therewith, and shall turn back and
cover that which cometh from thee."*

At the same time, humans had learned the benefit of adding manures to
agricultural soil and fish ponds in order to replenish nutrients depleted by
the perpetual removal of crops. Some of the so-called "night-soil" removed
from cities during the Middle Ages was destined for use on agricultural
land as fertilizer.

With the aggregation of populations into cities, intensified by the industrial revolution, terrifying outbreaks of Asiatic cholera, typhoid fever, and
other contagious diseases began to spread with a devastating toll of life.
Effects of these are matters of impressive historical record, unread and
unappreciated today in an era when modern water and sewage treatment
have limited waterborne disease in the developed world to occasional
outbreaks. However, in developing countries, little has changed.

The need for improved systems of water supply and waste disposal
was generally accepted only in the middle of the 19th century following
realization that the spread of Asiatic cholera (and, by implication, other
diseases) was caused by drinking water from wells infected with sewage
from diseased persons. Based on this discovery, major cities in technologically developed countries slowly began to construct one set of pipes to
convey water from an upstream site (some were already there) and another
to discharge wastes downstream (commonly leading into storm sewers that

were already there). In the process, Sir John Harington's invention of the water closet (flush toilet) came into practical use.[3]

Storm sewers, previously used to carry the erosive rush of rain water from city streets, became recipients of a rather dubious inheritance—municipal waste. With this the old adage, "rain to the river, sewage to the soil," went out the proverbial window. Wells disappeared from cities, and the once familiar cry of the water seller in the streets was heard no more.[4]

The greatest concern in terms of waste treatment in the 19th century was with the removal of disease-causing organisms and organic compounds from municipal and industrial wastes. This attack on organisms of disease needs no explanation now, although an explanation was needed then because of the poor state of medical knowledge. The reason for the accent on organic matter was that organic-rich wastes created unsightly appearances, foul odours, and fish mortality in natural waters because of oxygen depletion. The people who lived downstream and used the water for washing and drinking didn't like it, but only if they happened to be landowners was any improvement made.

Most of the treatment systems that have been developed for removal of organic matter from water depend upon the ability of microorganisms to break down organic compounds. In this process of biological oxidation, atoms of carbon, hydrogen, and nitrogen in organic compounds are converted into their oxides. Organic carbon is converted to carbon dioxide ($CO_2$); organic hydrogen to its oxide, water ($H_2O$); and organic nitrogen first to ammonium ($NH_4^+$) and then to nitrate ($NO_3^-$).

Biological oxidation is essentially a slow and controlled combustion of organic compounds, similar in principle to the more rapid process that takes place in the combustion of hydrocarbons in an internal combustion engine. Organisms are able to accomplish this controlled oxidation at low temperatures because of catalysts called enzymes. One can legitimately say that life depends on the trick of creating fire in water: combusting organic compounds at low temperatures and utilizing the energy released for growth and maintenance. Microorganisms in soils and water perform this service of oxidizing our wastes free of charge. Without this service, disease would spread, most higher organisms would be eliminated, and nutrients would not be recycled.

The first systems designed for microbial oxidation of organic matter in municipal wastes were sewage farms, in which the oxidizing ability of microorganisms in soil was used to advantage in agriculture. Lagoons, self-purifying sewage oxidation ponds, evolved somewhat later.

In the latter half of the 19th century "trickling filters" came into use. These are actually not filters, but thick beds of loose gravel designed for good aeration, through which sewage is flushed. Microbial "slimes," rich in bacteria, protozoans, and invertebrates, coat the gravel and carry out the oxidation. They do, however, remove silt, some pathogens, and protozoans.

The activated sludge system of sewage treatment was first described in 1913. It is a controlled engineering system, in which microbially rich ("activated") sludge that develops during treatment, is returned to the intake line so that organisms of decomposition can start to work immediately on incoming sewage. Methane produced from fermentation of the sludge is often used to help heat the treatment plant.

Ninety percent of the organic matter entering one of these continuous-flow systems is oxidized within 4–12 hours. In spite of their utility, systems of this sort in 1972 still served less than 15% of municipal waste generated in the United States and Canada. The typical "waste treatment plants" used by most communities in 1972 differed little from what was available in 1872—screens to remove coarse materials and a sewage lagoon to permit some organic decomposition to take place before wastes were discharged to streams. The situation is improving slowly. In 1999, the Organisation for Economic Co-operation and Development indicated that, in Canada, 48% of the population was served by activated sludge or other so-called "secondary sewage treatment." But 22% of the Canadian population still had no public sewage treatment at all. The figures for the United States are similar. In contrast, most countries of western Europe have from 85% to 98% of their populations served by sewage treatment.

Eutrophication happened in spite of sewage treatment for removal of organic compounds, because phosphorus and nitrogen are not appreciably removed during sewage treatment. As atoms, phosphorus and nitrogen are susceptible to recombination and rearrangement in molecules but not to destruction. Oxidation of organic matter during sewage treatment can even enhance the growth-stimulating effects of phosphates and nitrates by removing substances that inhibit plant growth, converting phosphorus and nitrogen to forms that are ideal for stimulating plant growth.

It took the better part of half a century for most sewage treatment engineers to learn that liberation of nutrients to water results in a "re-synthesis" of organic matter by algae in lakes and streams. One blessing is that, in synthesizing new organic compounds, photosynthetic plants release oxygen. It also took limnologists half a century to realize that they have an interest and a role to play in determining what goes on in sewage treatment plants. Fortunately, engineers and limnologists are now working more closely together. As a result, there are many modern advances in wastewater treatment that avoid some of the emerging problems with nutrients, toxic chemicals and pathogens.

Phosphorus removal, by precipitating phosphorus with iron or alum during sewage treatment, is now quite common. It is over 95% efficient in newer treatment plants. Following the discovery that chlorination of wastewater caused the formation of carcinogenic trihalomethanes by reaction with some organic substances, ozonation or ultraviolet (UV) treatment are now used in many areas. Ultraviolet raditaion has the advantage of destroying protozoans that cause waterborne illness, such as *Cryptosporidium*, which is quite resistant to chlorination. Recent advances indicate that UV also removes some of the pharmaceuticals, antibiotics, and personal care products that have recently been discovered to be common in wastewater and drinking water. However, if used to disinfect drinking water, UV and ozone treatment require that treated water must be transported and stored in secure ways, because there is not a "residual" disinfection effect as in properly chlorinated water. Proper control of chlorination also keeps the formation of trihalomethanes very low. Overall, the risk posed by chlorinated byproducts is very small compared with the high risk of illness from water contaminated with pathogens, and chlorination will probably remain the method of choice for water and wastewater disinfection for most areas for some time yet.

In the past few years, *reverse osmosis* has rapidly become the method of choice for water treatment in small communities, perhaps up to 2000 individuals, especially those with poor-quality water supplies. Water is forced under pressure through semi-permeable membranes with pores small enough to remove most of the chemicals. Although very high quality of water can be obtained without using chemicals, chlorine is usually added after treatment to avoid possible contamination in water distribution systems. The technique is rapidly becoming available to larger communities. In early 2007, a reverse osmosis plant is under construction in the Saddle Lake First Nation, Alberta, that will supply water to 7000 people.

The causes and consequences of human-caused eutrophication have now been exposed in general detail, tracing the course of history from the middle of the 19th century to the present. If one had to identify some of the main factors that caused nutrient concentrations to increase in waters during the 20th century, they would include:

1. the conversion of wells and outhouses into water mains, flush toilets, and sewage drains;
2. land conversion for human use, which stripped the catchments of freshwaters of their wetlands, riparian zones, and native vegetation and covered much of it with concrete and other impermeable materials;
3. greatly increased fertilizer production and use;
4. increased livestock culture has caused many problems analogous to those of human sewage; and
5. increased cost of treating wastes composed of more than 99.99% water.

Together factors 1–4 have caused cultural eutrophication. Alhough collection and treatment of wastes certainly improved sanitation, the discharge of sewage effluents to lakes and rivers accelerated the transport of nutrients to waterbodies, contributing to their rapid eutrophication. As well, the treatment of most livestock wastes is, unfortunately, still as rudimentary as human waste treatment a century ago. Obviously, the problem could have been prevented by ecologically based regulations that required nutrients to be intercepted. But we did not know this for much of the century. Now that we do, there is little excuse.

The overall design of human-dominated ecosystems has caused a one-way movement of nutrients from land to water. Nutrients are being usurped from soil by agricultural crops. The crops are then transported to areas of concentrated humans and livestock, where they are passed through intestines and kidneys, then delivered to streams, lakes, sediments, and the sea. The overall result of this "pathological togetherness" of humans and their domesticated animals is that nutrients once dispersed over large areas are now concentrated in a few watersheds, causing these to receive much more nitrogen and phosphorus than under natural conditions. As we have begun to understand these processes, progress in correcting them has been incremental. In general, governments still deal poorly with their responsibility for maintaining the environment to serve the common good.

As the need for greater production of food for the expanding human population continued to rise, nutrient disruption of aquatic ecosystems increased. Conversion of forests to grasslands, grasslands to fields, and culture of new, high protein varieties of grain have resulted in a greater flow of nitrogen and phosphorus fertilizers from land to sediments and the sea. We shall discuss this more in later chapters.

By 1970, the understanding of eutrophication had been extended to where little more could be done about the problem without more detailed understanding of the biological, chemical, and physical processes in lakes. In the latter part of this book, we shall describe key aspects of how these processes were discovered and developed, then describe how they have contributed to understanding of the eutrophication problem in the early 21st century.

# 3 | Lakes are Made of Water

*Deduction of the vertical circulation of water in lakes from a knowledge of the unique properties of water, temperature, density, and the influence of sun and wind*

## Water

WATER IS SUCH A COMMON SUBSTANCE that, like the air drawn into our lungs 30,000 times every day of our lives, we take it for granted. Pause for a moment to consider its significance.

Vented upwards from deep within the Earth through volcanoes, geysers, and fumaroles during over 5000 million years of geologic history, water has accumulated at the Earth's surface to such an extent that it now covers 70% of the surface to a mean depth of 3.8 kilometres, a little over 2 miles. Forests and deserts have been created in response to past alterations in the distribution and abundance of water precipitated from the atmosphere as rain. On several occasions during the past 2 million years, vast expanses of water in the form of glacial ice overrode continents, depressing the land, sculpturing valleys, depositing long mounds of gravel called moraines, damming lakes, re-charting river courses, and lowering the level of the world's oceans by more than 100 metres (330 feet).

As a medium of transport requiring neither roads nor rails, water has been a major factor in the development of trade and cultural exchange among diverse peoples of the Earth. On land, water means food and survival when it comes at the right times and places and famine when it does not. It provides electrical power for many technological conveniences in our homes and is the basis of our recreational pleasure in swimming, fishing, and boating. The thrill of cascading down a mountain slope on skis and the wild excitement of a hockey game also depend on water.

However, it is in the human body that water serves us best. Its high specific heat and circulation as the major component of blood permit the even distribution of temperature on which our lives depend. In summer, we are cooled by the high evaporative heat loss of water pumped outward to the body surface by a thousand infinitesimal factories of sweat glands in the skin. We could not survive without drinking it. It comprises two-thirds of our weight. It is the undisputed solvent for transport of food substances in the body and removal of wastes by the kidneys and lungs. As a cleansing agent, it prevents the spread of infections and disease.

In short, water is the chemical basis of life—according to many, perhaps a universal requirement for the origin and persistence of life. Our dependence on water has not been alleviated but has markedly increased with the advent of modern technology. This is illustrated by the high consumption of water in an average North American home, 380 litres (100 U.S. gallons) per person per day, equivalent in weight to 380 kilograms (836 pounds) per person per day. In Europe, where there is greater awareness of the value of water, and treated water is more expensive, consumption is only from one-quarter to one-half as great. In all cases, there is no other substance used to such an extent by humans.

Water is unevenly distributed in nature. Table 3.1 gives the percent distribution of the major forms of water on Earth that participate in the hydrologic cycle.

The fraction of water that occurs in the form of lakes and streams is minute. In spite of this, inland waters play disproportionately large roles in our lives, for we do not live within a volume of water but rather in proximity to water surfaces. Rivers have always been the avenues of commerce—at one time for produce, now more for waste. Lakes, on the other hand, are

| Form of water | Percentage |
|---|---|
| Oceans | 97.5 |
| Polar ice and glaciers | 1.7 |
| Groundwater | 0.75 |
| Freshwater lakes | 0.0075 (includes rivers) |
| Saline lakes | 0.008 |
| Soil and subsoil moisture | 0.005 |
| Atmospheric water vapour | 0.0009 |
| Rivers | 0.00009 |

valued for the peace and tranquility they confer on the mind. It is in the contrast of a lake that landscape achieves its fullest measure of beauty.

However, beneath the mirrored surface lies another design, more intricate in pattern: the wild mosaic of a folk dance performed by an infinite number of water molecules. Invisible forces holding the molecules together in this fluid architecture also act on a larger scale to endow lakes and even oceans with idiosyncrasies of their own, features so significant that no account of limnology would be complete without them. For this reason, it is important to understand how the chemical properties of water determine many aspects of the behaviour of water in lakes.

Water, a compound in which hydrogen and oxygen atoms are joined together in the ratio of 2:1, is not a simple substance. In fact, it was only as detailed knowledge of other chemicals accumulated over the centuries that the uniqueness of water came to be fully appreciated. If one had to predict its properties without ever having seen it, much as Mendeleyev in the 19th century accurately predicted the properties of chemical elements not known at the time, the expectation would be that water should exist as a gas at ordinary temperatures.

By analogy with hydrogen sulfide and other substances, the expected freezing point of water should lie at $-100°C$ ($-148°F$), and the boiling point would be at temperatures below the true freezing point ($0°C$, $32°F$). None of the unique attributes of water could be anticipated, such as the supremacy of its solvent power, its large heat capacity, and high surface tension. This contrast between expectation and fact signifies that, concealed within the structure of the liquid, there must be subtle forces binding water molecules together. How else could one explain the unexpectedly high amount of energy required to transfer molecules from the liquid state into vapour?

These hidden forces that operate on an infinitesimal scale within the magic liquid are called hydrogen bonds. Shifting about with more rapidity than flies before a swatter, they create flickering associations between one water molecule and another. This incessant "hand shaking" of molecules endows the liquid with a continuity and permanence in change. If our eyes could see the ballet of molecules in action, a quartet of dancers linked by hydrogen bonds (broken lines) might look like this:

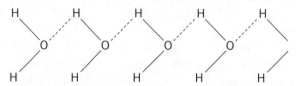

and an instant later like this:

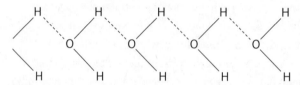

As Albert Szent-Gyorgyi once put, it, "Water is the only molecule that can turn around without turning itself around."

## Water Density as a Function of Temperature

AMONG THE SEVERAL ASPECTS of water that have fascinated chemists is one that, above all others, determines the pattern of the vertical circulation of water in lakes. This is the strange way in which the density of water changes with temperature. Of all liquids studied to date, only two have been found to exhibit maximum density at a temperature above, rather than at, the freezing point. One of these is water. The other is cesium chloride, a rare laboratory chemical that occurs as a liquid only at very high temperatures.

This peculiarity is matched by another unusual feature. Water differs from practically all other substances in being denser as a liquid than as a solid. As might be suspected, these two unusual features are related. They stem from a common cause.

Consider what an unusual place the world would be if ice were denser than water. Glaciers would sink to the bottom of the oceans, and ponds would freeze "under" instead of over. Oceans would perhaps be reduced to a thin film of supersaline water on top of a sedimentary rock that we would recognize as ice. Pressure of the atmosphere would be many times greater, laden with gases that would otherwise have passed to the sea. Probably life would never have evolved.

Ice has a density of only 0.92 grams per cubic centimetre as compared with approximately 1.0 gram per cubic centimetre for water in the liquid state. The low density of ice is a consequence of the way water molecules are arranged in the crystalline structure. Spaces are repeated at regular intervals as a result of fixed positions in which molecules are held in the solid. The more flexible arrangement of molecules in the liquid permits a denser packing.

Water in the liquid state exhibits its maximum density at a temperature of 4°C (39°F), just a few degrees above the freezing point. Two opposing forces are responsible for this unusual feature. One is the common property among all liquids and gases for the distances between molecules to increase with rising temperature. Considering this process alone, one would expect water to exhibit its maximum density at the freezing point, successively decreasing in density with rising temperature. The other force is the tendency for water molecules to group together in ice-like arrangements as the temperature of the liquid approaches the freezing point. Considering this second process alone, one would expect water to exhibit its minimum density at the freezing point, successively increasing in density with rising temperature. The combined operation of the two processes results in the closest overall packing of molecules at 4°C (39°F).

The most significant feature of the relationship between temperature and density of water is that it is not a straight-line relationship. The extent of density change per degree of temperature change depends on the actual temperature. For example, the difference in density between samples of water at 10 and 11°C is 12 times greater than between samples of water at 4 and 5°C—0.000095 grams per cubic centimetre versus 0.000008 grams per cubic centimetre. The difference in density between samples of water at 20 and 21°C is 26 times that of water at 4 and 5°C—0.000211 grams per cubic centimetre versus 0.000008 grams per cubic centimetre (Figure 3.1).

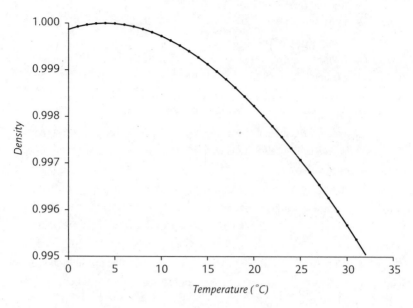

FIGURE 3.1: *Density of water as a function of temperature from 0 to 32 °C at a pressure of 1 atmosphere.*

The general rule is as follows: *the change in the density of water per degree change in temperature increases as the temperature departs from 4 °C, both above and below.*

This is an important rule; it is useful because it permits a number of characteristics common to all large waterbodies to be interrelated and explained in simple, fundamental terms. Knowing the rule, it is easy to understand why lakes become thermally stratified in the way they do, why dissolved oxygen tends to become depleted in the bottom waters of eutrophic (productive) lakes during summer, why lake trout disappear from lakes affected by cultural eutrophication, and how vertical circulation of water in tropical lakes can be restricted as a result of small gradients in temperature. Without the rule, these phenomena could only be interpreted as interesting, but isolated and unconnected, events.

The principal utility of the temperature–density rule is that it permits inferences to be made about whether the waters of a given lake are in complete circulation from measurements of temperature. *Thermal stratification implies incomplete vertical mixing of waters.* The logic is simple. Where there is a vertical gradient of temperature, there must also be a vertical gradient of density; and a vertical gradient of density is *prima facie* evidence of incomplete circulation.

## Thermal Stratification in Lakes

YOU MAY HAVE NOTED, when diving into a lake early in the season or on a particularly still and sunny day in summer, that water near the surface is often considerably warmer than the water below. The sun has warmed the surface waters more rapidly than the wind has mixed them with the underlying water. Layering persists because warm water is lighter (less dense) than cold water at temperatures above 4°C. Later in the season, when the sun-warmed waters have been mixed to greater depths, one can dive deeper without detecting any marked change in temperature. However, upon diving deeper still, one passes through a very sharp temperature gradient. The water below is nearly as cold as it was early in the season. This gradient is called *thermal stratification*, and the zone of maximum change in temperature with depth is known as the *thermocline*. If you have experienced such things, you may know a good deal more about lakes than you suspect.

In shallow lakes, well exposed to the wind, temperatures will be found to be practically constant from top to bottom—typically varying less than 2°C (3.6°F). This uniformity of temperature indicates that the waters are well mixed throughout.

In deeper lakes, three characteristic layers are present: (*i*) an upper zone of warm water in which temperature is more-or-less uniform throughout; (*ii*) an intermediate zone in which temperature declines rapidly with depth; and (*iii*) a lower zone of cold water in which temperature is again more-or-less uniform throughout. These three layers are termed *epilimnion*, *metalimnion*, and *hypolimnion*, respectively. The *thermocline* is defined as the point in the depth curve where the rate of temperature change is most rapid, which would be in the centre of the metalimnion in the diagram on the next page. They are depicted schematically in Figure 3.2.

The technical names for the three layers can easily be remembered from the meaning of the Greek roots from which they are derived. *Epi*, on or upon; m*eta*, among; *hypo*, under; and *limnion* from *limne*, pond or lake.

The metalimnion contains the *thermocline*, literally meaning *thermal gradient*. Although many limnologists use the terms metalimnion and thermocline interchangeably, the former is really a zone with a physical

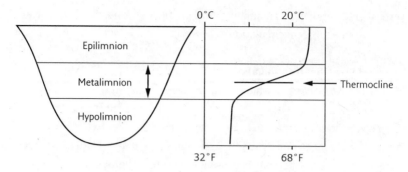

FIGURE 3.2: *Typical thermal structure of a deep lake at temperate latitudes in summer.*

volume, and the latter is the plane of maximum rate of change in temperature with depth.

The gradient of density in a thermally stratified body of water can be calculated from a knowledge of lake temperatures, using a conversion table listing the density of pure water for various temperatures, such as the one in G.E. Hutchinson's 1957 *Treatise on Limnology*, Volume 1. This procedure is tantamount to assuming the only factor regulating density in lakes is temperature. This is a reasonable assumption for most lakes. Exceptions are *meromictic* (partially circulating) lakes with bottom waters that are either salt laden, or in basins that are so deep and wind protected that they rarely, if ever, mix with the waters above.

The result of such calculations can easily be guessed. The gradient in density per unit depth in the epilimnion will be small or nil because of the uniformity of temperature. Exactly the same will be true of the hypolimnion. However, in the metalimnion, the gradient in density will rise to reach a maximum slightly above the point at which the rate of change of temperature with depth is maximal (Figure 3.3). In an early classical paper, E.A. Birge (1916) calculated the amount of energy that would be necessary to thoroughly mix a stratified lake.

The density gradient in the metalimnion forms a barrier that prevents the mixing of epilimnion and hypolimnion waters. Just as physical work must be performed in mixing cream and milk or oil and vinegar, so must physical work be performed in mixing layered water masses of different density. The only difference is that, in lakes, the mixed waters cannot separate again in the absence of energy, as do mixtures of aqueous and oily fluids. Other things equal, the amount of work required to mix layered

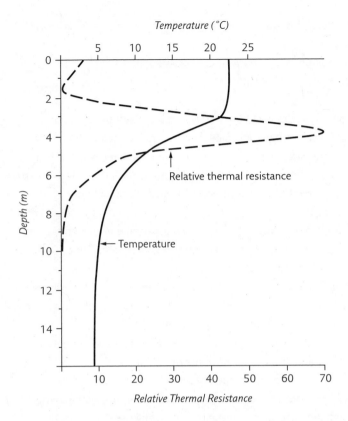

Temperature (°C)

Relative thermal resistance

Temperature

Depth (m)

Relative Thermal Resistance

FIGURE 3.3: *Temperature and relative thermal resistance (RTR) to mixing in Little Round Lake, Ontario, during summer. RTR is expressed as the difference in density between the top and bottom of each 0.5 m depth interval, calculated from temperature differences. RTR values have been multiplied by 105 for convenience in scales.*

Reprinted from Vallentyne (1974).

water masses of different density is proportional to the difference in density and the sharpness of the gradient.

Strange as it may seem from the minute changes in density associated with temperature changes in the thermocline, they are sufficient to prevent wind from completely mixing epilimnetic and hypolimnetic waters in lakes. Because of the marked reduction of turbulence and current velocities with depth, there is always some limit to the depth of the epilimnion created by even the strongest winds. A well-developed thermocline can be just as effective a barrier to downward moving currents as the shore or bottom of a lake. The characteristic change of temperature in the thermocline is thus a secondary reflection of the density barrier to mixing rather than a primary factor in itself.

Because temperatures in lakes can be measured more easily than density, limnologists typically infer the presence of a density barrier to mixing from temperature data alone. This practice is so common that students beginning the study of limnology find it difficult to understand why limnologists seem so preoccupied with the study of temperature. Of course, they aren't. It is the physical barrier to mixing that interests many of them. So, when a limnologist speaks of the physical "barrier" of the thermocline, remember—it is a barrier in density rather than temperature.

However, the thermocline is not only a physical barrier in the sense of restricting water movements; it also acts via temperature as a biological barrier affecting dispersal and growth of diverse aquatic organisms. Green plants and many forms of animal life—mayflies, caddisflies, molluscs, sunfish, and bass—reside in the warm, well-illuminated waters of the epilimnion in summer. Cold-water fishes such as trout, ciscoes, whitefish, and the small crustaceans, *Mysis* and *Pontoporeia*, on the other hand, rarely stray from the cool hypolimnion in summer. Other plants and animals find their requirements best satisfied in the metalimnion. Adopting a completely different way of life, some deep-water organisms, such as *Mysis*, make daily vertical migrations in the water column, coming to the surface at night and returning to deeper water during the day. Thus, the potential for growth and survival in different layers varies with the lifestyle of each species.

To understand how thermal characteristics of lakes change during the year, it is necessary to disentangle the separate influences of the sun and wind. This is a rather artificial procedure, because the two always act together in nature; however, in combination, the principles and ultimate causes involved are obscured. The procedure will be first to examine the characteristics of pure water in relation to the wavelengths of light emitted by the sun, later extending the logic to interpret events taking place in lakes.

### The Sun's Radiation

ONLY THREE PARTS of the sun's spectrum are of direct significance in the structure and metabolism of aquatic ecosystems. These are (*i*) ultraviolet (UV) radiation, consisting of wavelengths[1] shorter than those visible to the human eye; (*ii*) visible radiation; and (*iii*) infrared (IR) radiation, composed of wavelengths longer than those visible to the human eye. Properties of the radiation in each of these three regions are briefly described below.

## Ultraviolet Radiation

Some insects, aquatic crustaceans, and other organisms can detect UV light and have evolved complex behavioural patterns that are dependent on the UV radiation produced by the sun. On the other hand, humans are blind to UV light. We detect it only from tanning or sunburn reactions produced in our skin. Most of the UV radiation emitted by the sun is absorbed by the ozone layer in the upper atmosphere. The small amount reaching the Earth's surface is important in the production of vitamin D and in the behaviour of some organisms. UV light was once believed to be of little significance in lakes, but in the past 30 years, limnologists have discovered that high UV exposure can harm algae, bacteria, and small invertebrates; influence the outcome of competition between species; and cause other important effects. UV radiation is usually divided into UV-A, UV-B and UV-C. UV-A is solar radiation with wavelengths of 320–390 nanometres. UV-B is 280–320 nanometres, and UV-C is less than 280 nanometres. UV-B is the wavelength range that appears to be the most damaging to biological tissue, and it has been the primary focus of investigations of UV damage. UV-A actually activates tissue repair in some aquatic organisms. UV-C scarcely penetrates water and is of little consequence to aquatic ecosystems.

## Visible Radiation

The visible radiation emitted by the sun passes through the atmosphere with very little reduction in intensity except when the sky is covered with clouds. The colours we can discern within the visible spectrum are violet, blue, green, yellow, orange, and red, in order of increasing wavelength. Some violet and blue light at short wavelengths is scattered during passage through the atmosphere, in a plane at right angles to the direction of light from the sun. This scattering of blue light by small molecules and dust particles in the atmosphere is why the sky looks blue. For the same reason, sunsets look red. Because much of the blue light is scattered in other directions, red wavelengths are accentuated in the light that passes directly from the sun to our eyes. In wavelength, visible radiation is 390–700 nanometres.

*Infrared Radiation*

Also known as heat radiation, IR cannot be detected by the human eye. Owls, on the other hand, are able to detect IR radiation visually, a characteristic that enables them to locate mice in the dark from the heat radiated by their bodies. We sense IR radiation only by heat-sensitive nerve endings located in the skin. IR radiation is emitted by all hot bodies, of which the sun and an electric heater are two examples. It has wavelengths greater than 700 nanometres.

## Absorption of the Sun's Radiation in Lakes

OF THE TOTAL RADIATION emitted by the sun that reaches the Earth's surface, approximately half occurs as infrared radiation and half as visible radiation. Although a person with sunburn might justly claim otherwise, the amount of UV radiation is tiny, because much of it is absorbed in passing through the atmosphere, particularly the stratospheric ozone layer. The case is different on the moon. Lacking an atmosphere, the moon is constantly bombarded with UV radiation from the sun. For this reason, the suits and visors worn by astronauts have to be impermeable to UV radiation. Without such protection, the eyes and bodies of the astronauts would be burned. For similar reasons, the UV radiation falling on mountain ecosystems is greater than that falling on those at lower elevation, because the wavelengths pass through a thinner atmosphere at high altitude. For this reason, alpine lakes, with their clear waters, are of special concern, and many of the organisms in the lakes have protective pigments that reduce UV damage.

When radiation from the sun strikes a water surface, some is reflected, some is absorbed and in the process converted into heat energy, and some is transmitted through the water column depending on its thickness and the wavelengths of light involved. As an example of the importance of wavelength, place a hand on a dark-coloured car sitting in the sun, then on a light-coloured car. The dark car will always be much warmer, because it absorbs more of the wavelengths emitted by the sun. In contrast, the light-coloured car reflects much of the solar energy that hits it.

Water selectively absorbs some wavelengths of light more than others. IR, for example, is absorbed more strongly by water than visible radiation.[2] Likewise, in pure water, the transmission of visible light in a given water column increases continuously from the red part of the spectrum down to the blue part of the spectrum. At still shorter wavelengths, the situation

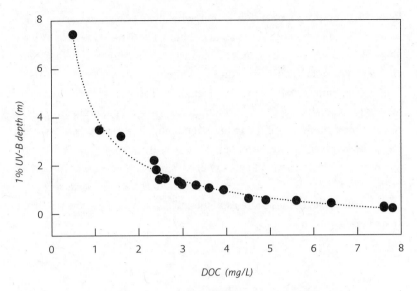

FIGURE 3.4: *The relationship between dissolved organic carbon (DOC) and the penetration of ultraviolet radiation in north temperate lakes.* From Schindler et al. (1996a).

reverses. UV light, violet light, and blue light at shorter wavelength are absorbed to a greater extent than blue light. Although it might at first be thought that water often looks blue because of its transparency to blue light, the more common reason is because of reflected blue light from the sky.

Under water, other factors modify penetration of the sun's radiation. Suspended silts and clays, dissolved humic material (generally measured as dissolved organic carbon, or DOC), and microscopic plant life can all exert a pronounced effect, decreasing the transmission of light and shifting the maximum transmission toward longer wavelengths. UV is rapidly absorbed by coloured dissolved organic matter, which acts as a natural "sunscreen" for lakes (Figure 3.4). In clearwater lakes, significant UV can penetrate to depths of several metres, but in lakes with several milligrams per litre of dissolved organic carbon, UV wavelengths can be absorbed within a few centimetres of the surface, which is fortunate, because at high intensities, it can be extremely damaging to phytoplankton and small animals. Thus, coloured dissolved organic matter can act as a "natural sunscreen" to protect aquatic organisms.

A useful rule known as Lambert's Law, concerning the absorption of light by homogeneous liquids and solids, was formulated by J.H. Lambert in the 18th century. Lambert found that, when light of a given wavelength passed through a homogeneous solid or liquid, the percentage of light absorbed by each successive unit of distance was constant. Figure 3.5 shows an example. If 90% is absorbed by a column of water 1 metre in length, a column of water 2 metres in length will absorb 99%, of the incident light. (Ninety percent of the original 100 units of light is absorbed by the first metre and 90% of the 10 units remaining is absorbed by the second metre. This leaves only one unit of light to pass through the entire 2-metre column.) The depth to which light penetrates depends on the clarity of the water, but it is easy to predict for well-mixed water masses, because it is attenuated at a logarithmic rate as in the following equation, which represents Lambert's Law:

$$I_z = I_o e^{-kz}$$

where $I_z$ is the irradiance at depth $z$, $I_o$ is the irradiance at the surface, $k$ is the extinction coefficient, and e is the base of natural logarithms.

Because water does not evenly absorb light at different wavelengths, light of different wavelengths is attenuated at different rates, and for natural daylight, the extinction coefficient would be a composite of these based on the sun's spectrum and the colour of dissolved and suspended matter in the water. For similar reasons, the colour of a Secchi disc or other white object suspended in a lake is affected. In pure water, the white object would appear blue at depth. In our clearest lakes, a Secchi disc commonly appears blue-green. In lakes with higher concentrations of plankton or dissolved organic carbon, it can appear green, yellow, or even red-orange, depending on the wavelengths of maximum transmission (Figure 3.6). In general, the overall penetration of light decreases as more dissolved organic carbon, plankton, or silt are added.

Knowing that (i) light energy is converted into heat energy on absorption, and (ii) the greatest amount of absorption must occur closest to the light source, a simple prediction can be made for lakes. Lakes should be most intensively heated by the sun in the uppermost layers where greatest absorption occurs. If a lake could be kept under completely quiescent conditions in continuous sunlight, temperatures would be highest at the surface and decline rapidly (and, thanks to Lambert's Law, exponentially) with depth. The most dramatic changes in temperature would occur immediately below the water surface.

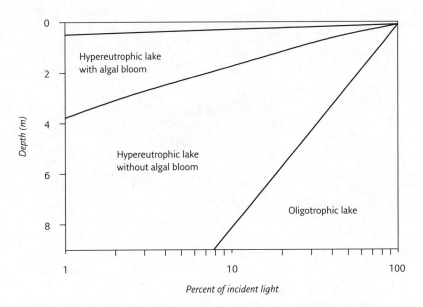

FIGURE 3.5: *Attenuation of light with depth in water, demonstrating the logarithmic decrease with depth described by Lambert's Law.*

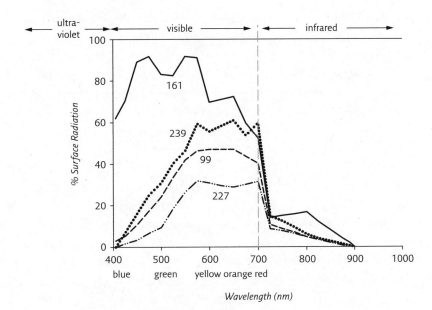

FIGURE 3.6: *The spectral shift in wavelengths transmitted through water that occurs at depth with increasing concentrations of DOC. Lake 161 is a very clear lake with DOC about 2 mg/L. Maximum transmission is in blue and green wavelengths. Lake 227 had over three times the concentration of DOC in Lake 161, with maximum transmission at orange and red wavelengths. The other lakes are intermediate. Wavelengths are in nanometers. Wavelengths over 700 nm are in the infrared range.* After Schindler (1971a).

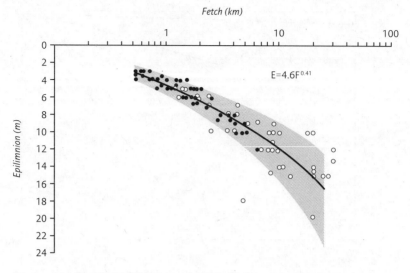

FIGURE 3.7: *The relationship between the fetch of north temperate zone lakes and epilimnion depth.* From Patalas (1984). Printed with permission from E. Schweizerbart'sche Verlagsbuchhandlung (Naegele u. Obermiller) Science Publishers (http://www.schweizerbart.de).

This prediction can easily be tested in a simple home experiment.[3] An IR lamp or high-intensity light bulb, corresponding to the sun, is positioned over an aquarium filled with cold water from the tap. After an hour or so an ordinary outdoor thermometer can be used to show, as predicted, that the depth distribution of temperature is exactly as described above. Temperatures will be highest at the surface and decline rapidly with depth. The major work of the sun is confined to the uppermost few metres of water, where practically all the IR and visible radiation of the sun is converted into heat.

The situation in natural lakes is entirely different. Waters of the epilimnion are uniform in temperature to considerable depths, not uncommonly 3–20 metres (10–65 feet), depending on the size of the waterbody. Wind is the primary agent responsible for distributing the sun's energy to heat deeper parts of lakes. The transfer is achieved by means of waves and water currents. For lakes in continental areas away from the moderating effects of the oceans, the depth of the thermocline is well predicted by the size of a lake, with the maximum fetch being the best predictor (Figure 3.7). The difference between the results obtained in the aquarium experiment and those observed in natural lakes is principally due to the action of the wind.

In the case of very turbid waters, rich in suspended clays, or brownwater lakes rich in humic matter, almost all the sun's radiation may be trapped in the uppermost few centimetres (inches). On the other hand, in extremely clear waters (such as Lake Tahoe on the California–Nevada boundary; Crater Lake, Oregon; Great Bear Lake in northern Canada; or lakes in high alpine areas), blue light from the sun may penetrate to depths in excess of 100 metres (330 feet).

## Temperatures of Deep Waters

A STRANGE SITUATION OCCURS in a class of lakes known as partially circulating or *meromictic* lakes. Water temperatures in the deepest parts of these lakes are sometimes *higher* than in the immediately overlying water. The circumstance that permits this is the occurrence of high concentrations of dissolved chemicals in the deepest waters.[4] The high density caused by these dissolved substances creates a year-round barrier to mixing that is just as effective as the barrier of the thermocline in summer. Depending on the depth and transparency of the overlying water, the salt-laden bottom water tends to accumulate heat like a closed car parked in the sun. The heat is easily gained through absorption, but lost only with difficulty. The most extreme case known is Hot Lake, in the State of Washington, USA. Because of its shallowness—the lake is only 3 metres (10 feet) deep—the temperature of the salt-rich bottom-water rises to 50°C (122°F) in summer and can be as high as 30°C (86°F) in winter under a cover of ice.

The temperatures of bottom waters in deep temperate lakes are typically only 4–5°C, with surface waters in temperate regions reaching 20–25°C in midsummer. Bottom waters in tropical lakes generally lie in the range of 15–25°C (59–77°F). Surface temperatures are typically 5–10°C (9–18°F) higher. Although temperature differences between the epilimnion and hypolimnion are greater in temperate lakes, the difference in density stratification is smaller than in some tropical waters. This is because of the large density differences per degree of temperature change at high temperatures (Figure 3.1). A good general rule to remember is bottom temperatures of lakes that do not freeze over are rarely lower than the mean temperature for the coolest month of the year. This stands to reason, because air temperatures have to be low for a long time in order to cool what usually amounts to a large volume of water.

Although the thermocline retains its structure throughout spring and summer, it is often tilted from a horizontal position when a prolonged

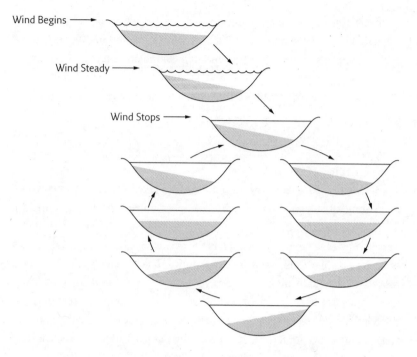

Wind Begins ⟶

Wind Steady ⟶

Wind Stops ⟶

FIGURE 3.8: *The depiction of an internal seiche, the rocking of the thermocline that results from wind blowing the less-dense epilimnion water to the downwind side of a lake. The rocking can continue for days after the wind ceases to blow, because of very low friction in the thermocline.* Reprinted from Vallentyne (1974).

wind from one direction piles up epilimnetic water at the downwind end of a lake. This phenomenon was described by E.A Birge in 1910 and thoroughly studied by Clifford Mortimer, beginning in the early 1950s. In extreme cases, the depth of the thermocline can differ by several metres at the upwind and downwind ends of a lake. The angle of tilt can be great enough that the thermocline rises to the surface at the upwind shore. When the wind ceases the situation is unstable with respect to gravity. The thermocline then begins to ponderously oscillate back and forth in teeter-totter fashion until damped out by opposing wind stresses or frictional losses. These motions, known as *internal seiches*, induce turbulence into the hypolimnion and the sediments at the depths affected by the seiche (Figure 3.8). They are important for a number of other reasons.[5] You may have noticed, as have the citizens of Toronto on occasion, peculiar relations between temperatures of air and water at bathing beaches. Sometimes the water is unseasonably cold when the sun is shining and the air is hot. At other times, the reverse is true. When such things happen, they are usually the consequence of an internal seiche from a previous wind.

A simple and yet effective model of an internal seiche can be made by reversing the normal procedure for mixing ice, whiskey, and water in a glass. Fill a wide glass about one-third with water. Tilting the glass at an angle, slowly add whiskey to taste, layering it carefully on top of the water. The upper layer of whiskey corresponds to the epilimnion and the lower layer of water to the hypolimnion. There is a gradient of density between them corresponding to the density gradient associated with the thermocline. Tilt the glass back and forth two or three times attempting to move in unison with the motion of the whiskey. Then hold the glass steady in a horizontal position to observe internal seiches. Finally, add ice as desired and enjoy the drink. (For non-drinkers, a similar experiment can be performed by layering water over sugary tea).

## Seasonal Changes

SEASONAL CHANGES of temperature and density in lakes are best described using as an example a temperate lake that freezes over in winter. When ice coats the surface of a lake, it is said to be "inversely" stratified with respect to temperature. Cold water at $0°C$ lies in contact with ice, and warmer water, between 0 and $4°C$, occurs below.

With the coming of spring, ice melts, and the waters are mixed by wind. Shortly, the lake is in full circulation, and temperatures are approximately uniform throughout (close to $4°C$). With further heating from the sun and mixing by the wind, the typical pattern of summer stratification develops, dividing the lake in epilimnion, metalimnion, and hypolimnion.

In late summer and early autumn, as the lake cools in sympathy with its surroundings, convection currents of cold water are formed at night when air temperatures are cooler than surface water. These sink to find their appropriate density level, mixing with warmer water on their way down. With further cooling, and turbulence created by the wind, the thermocline moves deeper and deeper. The temperature of the epilimnion gradually approaches that of the hypolimnion. Finally the density gradient associated with the thermocline becomes so weak that it ceases to be an effective barrier to downward moving currents. The lake then becomes uniform in temperature indicating it is again in full circulation. With still further cooling, ice forms at the surface to complete the annual cycle.

With this, the brief interlude with water comes to an end. In relating the infinitesimal forces within the magic liquid to the properties of lakes, some of the mystery of water has been both revealed and lost. Events perhaps unrealized by the casual observer have been explored in principle and detail, permitting an appreciation of how sun and wind, acting in unison, determine many of the thermal characteristics of lakes. Looking at the matter in another way, perhaps one could equally well say that these great powers are merely acting out their parts in a play that has been written within the confines of a single water molecule.

# 4 | How Lakes Breathe

*Factors that regulate the balance of carbon dioxide, oxygen, and organic matter in lakes; the exchange of dissolved gases between lakes and the atmosphere; and the effect of eutrophication on both processes*

THE CIRCULAR MOVEMENT of hands on the face of a clock, the migration of air in and out of the lungs, the transition from morning stillness into an afternoon breeze, the up and down motion of pistons in a car engine, the evaporation and precipitation of water on Earth, and the succession of life in the biosphere—all of these cyclic processes "run," only because they are driven by energy from the sun. Travelling at a velocity of 300,000 kilometres (186,000 miles) per second, this perpetual flow of radiation from the sun consumes only 8⅓ minutes in its passage to Earth. On arrival, most is converted directly into heat. Only a minor fraction is transformed into stored forms of energy that persist for varying lengths of time, such as plants or fossil fuels, eventually decaying into uniformly distributed heat.

Modern technology could never have advanced to its present state without this stored energy of the sun. The calories in our daily bread are derived from the sun. Electric lights powered by water are really powered by the sun's energy that, days, weeks or millennia before, was consumed in evaporating water from a neighbouring sea, transporting it to a higher elevation on land. With a greater time lag, the combustion of coal, natural gas, and petroleum releases energy that fell on the Earth as sunshine and was stored by photosynthesis tens to hundreds of millions of years ago. No modern technologist would ever think to question why our ancestors worshipped the sun, yet somehow in the luxury of modern technology, many people seem unaware of our dependence on the sun. Only in a child's instinctive bedtime plea for the hall light to be left on and the door ajar do we recall our deep inborn fear of being alone in the dark.

The sun is the source of virtually all utilizable energy at the Earth's surface. One hundred and fifty million kilometres (93 million miles) away, less than 1/2,000,000,000th of its radiation is intercepted by the Earth. Of this, only a small proportion (perhaps about 1/1000th on the average) is annually converted photosynthetically into new plant growth.

## Photosynthesis and Respiration

IN THE COURSE OF PHOTOSYNTHESIS, green plants use the sun's energy to manufacture energy-rich organic compounds from carbon dioxide and water. Oxygen is released as a by-product. In the reverse process of cellular respiration, plants and animals use oxygen to convert organic carbon compounds back into carbon dioxide and water, releasing the chemically stored energy of the sun for metabolism and growth.[1] The unified action of life as a whole depends on a favourable relationship between these two major processes of photosynthesis and respiration. Upsets of this balance in natural waters affect every form of aquatic life from the smallest bacteria and algae to insect larvae and fish. So it is that a lake, and even the biosphere, can truly be said to breathe.

Chemical and biological transformations involved in the regulation of this "breathing" are exceedingly complex, but the basic principles can be outlined without becoming entangled in detail. It is important that we understand these basic principles. We are living at a time in history when humans' ability to change the biosphere is both immense and real. For example, scientists estimate that we are now transforming about 40% of the Earth's photosynthesis to human use, and we have accelerated the natural cycles of phosphorus, nitrogen, and many other elements by twofold or more. If we should err environmentally in our societal actions, it will probably be because we do not fully understand how ecosystems operate in nature or how seemingly minor changes on our part could perhaps induce events beyond our control. When such things happen (and they are happening on local and increasingly larger scales each day), it is perhaps time to pause and reflect on how far we should proceed with modification of the Earth when our understanding of how ecosystems operate is so rudimentary.

In the continual interplay and adjustment between living and nonliving components of the biosphere, changes in one part of the system have induced reciprocal changes in the other. Ecosystems have not remained static in time; like most other things in nature, they have evolved. To cite one past example of this evolutionary interplay, it is generally accepted that the entire supply of oxygen in the Earth's atmosphere has been derived from the activities of photosynthetic plants. As a consequence of this single event, the world is now a vastly different place than it would otherwise have been. The rate of geological erosion, the chemical composition of natural waters, and the final products of rock weathering have all changed dramatically as a result of the addition of oxygen to the atmosphere. More significantly, neither humans nor any animals on Earth today could breathe were it not for this inheritance from organisms that we often refer to as the "lower forms of life." Therefore, a discussion of how the biosphere began to "breathe" may form a fitting prelude to the description of the breathing that takes place in lakes.

To the ancients, life consisted of ordinary matter endowed with "vital spirits." In their ignorance of events that take place in microscopic dimensions, they were convinced that worms and maggots were generated "spontaneously" in concoctions such as cow dung and soil. Did they not see it happen with their own eyes? Even in the 17th century, van Helmont, one of the more respected scientists of his time, believed a recipe of wheat grains and a dirty shirt would, if incubated for 21 days, spontaneously generate mice. He was only surprised by their exact resemblance to mice of natural birth.

With the improved vision following the invention of the microscope and the experiments of Redi, Spallanzani, and Pasteur, it was eventually recognized that all life originated from pre-existing life—as some put it, all eggs from an egg. With that recognition, the scientific question of the primordial origin of life came into being.

Little progress has been made in answering the question beyond the demonstration that a "soup" of organic chemicals was probably present on Earth prior to the appearance of life. This "soup" is thought to have existed at a time when the atmosphere was devoid of oxygen and when ultraviolet light penetrated to the surface of the Earth. Experiments have shown that many chemical building blocks of life are created when some of the simple gases presumed to have been present in the Earth's primitive atmosphere are irradiated with ultraviolet light.

It is believed that, through a slow process of chemical evolution, molecules aggregated into primitive living things that fed on the chemical energy of this preformed "soup." Only as the "soup" became depleted in constituents essential for growth did some aberrant forms develop the ability to synthesize the missing molecules, as the latter disappeared from the environment, one by one. Eventually photosynthetic plants appeared that could make their own "soup" in a cell. Organisms unable to do this either died out or became dependent on photosynthetic plants.

The development of plant photosynthesis was perhaps the most profound event in the biological history of the Earth. It permitted the utilization of abundant substances in nature as raw materials for photosynthesis; it created a supply of oxygen as a "fuel," which in turn permitted the later development of higher forms of plant and animal life; and ozone was produced from oxygen in the upper atmosphere, shielding the surface of the Earth from biologically harmful ultraviolet light. It is fascinating to consider that the events paving the way for our existence today started with primitive photosynthetic cells several billion years ago.

As mentioned earlier, oxygen is a byproduct of photosynthesis. While some of the oxygen produced is re-used by plants, photosynthesis has exceeded respiration for much of the Earth's history, allowing oxygen to accumulate in the atmosphere, where it is available to other organisms. The utilization of oxygen by non-photosynthetic forms of life is a striking illustration of how the biosphere recycles elements, using them to advantage. The basic conversions that happen in photosynthesis and respiration are illustrated in the equation below, where $CH_2O$ is "shorthand" for organic matter:

$$E + CO_2 + H_2O \underset{respiration}{\overset{photosynthesis}{\rightleftarrows}} CH_2O\downarrow + O_2\uparrow$$

| E | | $CO_2$ | | $H_2O$ | | $CH_2O\downarrow$ | | $O_2\uparrow$ |
|---|---|---|---|---|---|---|---|---|
| solar energy | + | carbon dioxide | + | water | | organic matter | + | oxygen |

This equation shows that, in photosynthesis, carbon dioxide and water react to form carbohydrates that store energy and oxygen. In respiration the process is reversed, with the release of energy.

A more complex equation is often used to include nitrogen and phosphorus in the calculation:

$$106CO_2 + 16NO_3^- + HPO_4^- + 122H_2O + 18H^+ \rightleftarrows (CH_2O)_{106}(NH_3)_{16}(H_3PO_4) + 138O_2$$

This equation depicts carbon, nitrogen, and phosphorus in the ratios observed in typical marine organic matter. This ratio is known as the *Redfield ratio* after Alfred Redfield, who first noticed that it was generally applicable to phytoplankton in the world's oceans. Although the ratio does not apply strictly to all organisms, it is useful for rule-of-thumb calculations about the ratios of nutrients necessary to form organic matter. Later, we will use this formula to deduce situations where phosphorus or nitrogen is likely to limit algal production, and where nitrogen-fixing Cyanobacteria are likely to dominate the phytoplankton because they can rely on atmospheric $N_2$ to make up for any deficiencies in the supply of dissolved ionic forms of nitrogen, giving them an advantage over other taxa.

If there had been a perfect balance between photosynthesis and respiration during the history of the Earth, oxygen could never have accumulated in the atmosphere. It accumulated because not all organic matter formed by photosynthesis was oxidized back to carbon dioxide and water. Some was deposited in sediments, forming the fossil organic carbon compounds now occurring in black shales and other sedimentary rocks, coal, petroleum, oil sands, and natural gas. Some carbon is stored for thousands of years in peat bogs and lake sediments and for hundreds of years in trees and soils. For every atom of carbon fossilized or stored in biological material, a molecule of oxygen passed into the atmosphere.

In the past, some scientists claimed that humans might deplete the supply of oxygen in the atmosphere as a result of oxygen consumed in the combustion of fossil fuels, disturbing a balance that it has taken nature several billion years to create. However, recent studies show that the oxygen balance is not likely to be disturbed, because most organic carbon does not occur in a form that is economical to mine or burn.

Other scientists have more realistically contended that increased levels of carbon dioxide in air brought about by the combustion of fossil fuels could raise the temperature of the Earth. This suggestion is based on the "greenhouse" effect. Although solar energy of most wavelengths strikes the Earth, back-radiation is largely as infrared radiation. Carbon dioxide absorbs some of the infrared energy the Earth radiates to space, causing the Earth to warm. The immense buffering systems of oceans and photosynthetic plants will eventually handle this matter with ease, particularly photosynthetic plants, because, for every molecule of carbon dioxide assimilated a molecule of oxygen is released. However, we now know that the transfer of $CO_2$ to oceans and plants is a slow process and that the complete removal of the excess $CO_2$ released by human activity will require several centuries. It is during this non-equilibrium period that there is great concern for the transformation of ecosystems by climate warming and for the disruption of human societies by shortages of water and food.

Although the exact times of the origins of life and photosynthesis on Earth are imperfectly known, it is certain from well-preserved remains of algae in ancient deposits on the north shore of Lake Superior that both events must have occurred more than 2.5 billion years ago. More problematic biological remains are known from South African rocks, 3.3 billion years old. Lack of deposits of greater antiquity leaves a question mark between then and the time the Earth was formed, about 4.6 billion years ago.

## How Lakes Breathe
TURNING NOW to the question of how lakes "breathe," carbon dioxide deserves first attention because of its importance in photosynthesis. Among the 15–20 chemical elements required for the growth of different species of aquatic plants, carbon is of prime significance because it comprises approximately half the weight of organic material in living cells. All living organisms have a high demand for carbon, but as discussed in connection with eutrophication, carbon only rarely limits the growth of aquatic plants in inland waters. In practically all cases (excluding some enriched with sewage wastes and some bottom-living algae where reduced circulation of water restricts the supply of inorganic carbon), the environmental supply of carbon is more than adequate to meet the demands of photosynthetic plants. The physical and biogeochemical mechanisms that must operate to get this carbon to the locations where it is available to aquatic plants are described more thoroughly in Chapter 9.

Practically all chemical elements known to occur on Earth can be detected in most inland waters. For some elements, such as uranium and gold, the concentrations are so minute (parts per trillion or less) that special methods and precautions are necessary in conducting analyses. Calcium and carbon, on the other hand, usually occur in sufficiently high concentrations (parts per million or more) to permit detection by routine chemical analysis. Most inland waters are dominated by calcium and bicarbonate, simplifying some of the more important processes that we must understand to know how the chemistry and the flora and fauna of lakes interact.

Within their drainage basins, river courses act as collecting ducts for precipitation from the atmosphere, the products of rock weathering and vegetational synthesis on land, and the wastes of humans and other animals. In granitic areas with thin soil, such as the vast Precambrian Shield areas of northern Canada, Scandinavia, and northern Russia, the amount of inorganic material leached from the land is low. Lakes and streams in these areas accordingly have low concentrations of dissolved inorganic salts and are said to contain "soft" water. In contrast, lakes and streams in regions of sedimentary rock drainage typically have high concentrations of dissolved inorganic salts. They are referred to as "hard" waters when they contain appreciable concentrations of calcium and magnesium salts. Water hardness is mainly due to high concentrations of calcium, magnesium, and other doubly and triply positively charged ions that form precipitates with soaps, which can cause problems in using soaps for washing clothes if water is too hard.

Atmospheric gases are also dissolved in natural waters. Surface waters in intimate contact with air typically contain 15–25 parts per million (ppm) of dissolved nitrogen by weight and 8–14 parts per million of dissolved oxygen by weight. (Nitrogen is actually less soluble than oxygen in water under conditions of equal gaseous pressure. The higher concentration of nitrogen in natural waters is due to the fact that its partial pressure in the atmosphere is more than three times that of oxygen.) When these gases dissolve in water, they do not change in chemical structure. In contrast, when carbon dioxide dissolves in water, it reacts chemically with water to form a new compound called carbonic acid. Carbonic acid, in turn, exists in equilibrium with two other forms of inorganic carbon: bicarbonate and carbonate.

### Inorganic Carbon

IN ALL, five different forms of inorganic carbon occur in natural waters: carbon dioxide, carbonic acid, bicarbonate, carbonate, and calcium carbonate. They are all readily interconvertible depending on environmental conditions. Addition of acid to a sample of natural water causes a shift in the forms of inorganic carbon toward carbon dioxide. Addition of base causes a shift of inorganic carbon toward carbonate. If concentrations of carbonate and calcium get too high, a precipitate of calcium carbonate can form. (Caustic soda and lyes are bases. Acids and bases neutralize each other.)

Conversion of inorganic carbon from one form to another is very much like exchanging the currencies of different countries as one travels. The only major point of difference is that, in the chemical system, the exchange has to be performed in a defined sequence. Nature does not impose a charge in the form of energy for making the exchange.

The situation is entirely different with respect to the interconversion of inorganic and organic forms of carbon. Nature imposes a heavy charge for converting carbon dioxide into organic carbon compounds. The fee, in the case of photosynthesis, is energy, paid by the sun. Part of the fee is returned in respiration when organic compounds are broken down into carbon dioxide and water.

Another part of the organic carbon is released into water. This joins dissolved organic matter that leaches into lakes from soils in forests, grasslands, and especially wetlands in their catchments. The terrestrial component is usually highly tea-coloured. Together, the *allochthonous* (produced by photosynthesis outside the lake) and *autochthonous* (produced by photosynthesis within a lake) sources form what is known to limnologists as dissolved organic matter (DOM). Dissolved organic matter is usually measured by its carbon content, so that it is frequently referred to as dissolved organic carbon (DOC). Dissolved organic matter is fuel for the microbial community of lakes, often contributing greatly to respiration of $CO_2$ and to the energy available to aquatic protozoa and zooplankton.

The futility of studying carbon dioxide without reference to the other forms of carbon and other chemicals in aquatic ecosystems should be apparent. When one form of inorganic carbon is removed from or added to the system, the others automatically readjust themselves to minimize the effect of the change. Photosynthesis and respiration also affect the forms and abundance of inorganic carbon in natural waters. In summary, the entire system of carbon compounds in nature has to be examined in the context of the cycle of carbon and many other chemicals.

Three concepts are necessary as a prelude to understanding how the carbon cycle operates in nature. These are ions, pH, and buffers and buffering capacity,

## Ions

An ion is an electrically charged atom or molecule. A variety of different kinds of ions occur in natural waters. Many of these enter into the composition of our bodies and are essential for health and survival. Some ions carry a positive charge, such as calcium ions ($Ca^{2+}$) and hydrogen ions ($H^+$). They are known as *cations*. Other ions carry a negative charge—bicarbonate ions ($HCO_3^-$), carbonate ions ($CO_3^{2-}$) and hydroxyl ions ($OH^-$). These are known as *anions*. In water, the sum of cations must equal the sum of anions, to maintain electroneutrality, one of the basic principles of water chemistry. Hydrogen ions and hydroxyl ions unite to form water ($H^+ + OH^- = H_2O$).

## pH

The term pH refers to the concentration of hydrogen ions on a negative logarithmic scale extending from 0 (acidic) to 14 (basic). The lower the pH the higher the concentration of hydrogen ions and the more acidic the solution. A decrease of one unit in pH corresponds to a 10-fold increase in the concentration of hydrogen ions. For example, at a pH of 5, hydrogen ions are 10 times more abundant than at a pH of 6.

## Buffers and Buffering Capacity

Buffers stabilize pH by neutralizing hydrogen ions and hydroxyl ions. Bicarbonate is the principal buffer in natural waters, in the blood of animals, and in the fizzy drinks people often take "the morning after." Soft waters tend to be poorly buffered; hard waters, well buffered. Water, by itself, has no buffering capacity.

Two "kitchen" experiments may help to cement these concepts together and relate them to the carbon cycle. For the first experiment, dissolve a teaspoonful of baking soda (sodium bicarbonate) in a half glass of water. Next, slowly add some vinegar (acetic acid) in small amounts. The baking soda solution fizzes vigorously on addition of vinegar.

The explanation of this experiment is as follows. The solution of baking soda in water contains sodium ions and bicarbonate ions. Vinegar acts as a source of hydrogen ions that convert bicarbonate ions into carbonic acid and carbon dioxide. Carbon dioxide, coming out of solution, causes the effervescence. A determination of pH before and after the addition of vinegar would show no appreciable change for the first few additions. The solution was buffered by the bicarbonate present in baking soda. However, as more vinegar is added, the supply of bicarbonate ions will be exhausted, after which the fizzing will stop and the pH of the solution will decrease rapidly.

The second home experiment will have already been performed if you live in an area with moderately hard water. Locate a kettle that has been in use for some time. You will probably find that it is coated on the inside with a deposit of calcium carbonate, also known as "boiler scale." Alternatively, simply fill a pot with water and evaporate it to near dryness. The whitish deposit of calcium carbonate forms as a result of the removal of carbon dioxide from water on heating, causing the water to become supersaturated with calcium carbonate.

The explanation is as follows. The solubility of all gases in water decreases with increasing temperature. The feature peculiar to carbon dioxide in natural waters is that for each molecule removed, a hydrogen ion and a bicarbonate ion react to form carbonic acid; and a hydroxyl ion and bicarbonate ion react to form a carbonate ion and a water molecule. The net result is the formation of a carbonate ion for each molecule of $CO_2$ removed. In waters containing high concentrations of calcium ions (hard waters), excess carbonate ions may precipitate out as calcium carbonate.

A phenomenon similar to the formation of "boiler scale" occurs naturally in some hard-water lakes as a result of removal of $CO_2$ by photosynthesis and the warming of water in summer. Both of these processes make calcium carbonate less soluble, so that it precipitates. A light milkiness appears in the water resulting from the formation of small crystals of calcium carbonate. In some lakes, these periods are conspicuous enough to be called "whitings." Calcium carbonate is also precipitated on surfaces of pondweeds and filamentous algae growing on rocks near the shore as a result of the removal of carbon dioxide in photosynthesis. These white precipitates are often referred to as *marl*. Depending on the concentrations of other chemical elements, the marl can contain magnesium carbonates and traces of other salts, as well as some organic matter.

## Carbon Dynamics as a Mechanical Bird

THE VARIOUS FACTORS affecting the forms and abundance of inorganic carbon in lakes can be summarized by means of the mechanical bird shown in Figure 4.1. The chemical equations show how the several forms of inorganic carbon are interrelated, the dots indicating relative abundance of carbon atoms in different forms. For a given environmental change, the flow of carbon atoms in the system follows the law of gravity and is determined by the angle of tilt on the bird's wings. Escape of carbon atoms from the wings indicates passage of carbon dioxide from water into air or precipitation of calcium carbonate from solution. The "left turn" position of the mechanical bird shows what happens when acid industrial wastes enter natural waters: the pH drops, carbon atoms flow in the direction of carbon dioxide, and part of the carbon dioxide is expelled to air. The "right turn" position shows what happens when basic industrial wastes enter natural waters: the pH rises, carbon atoms flow in the direction of carbonate, and some of the carbonate ions are precipitated out of solution as calcium carbonate if the waters are sufficiently hard.

The "wings down" position of the mechanical bird corresponds to removal of carbon dioxide from water due to warming or photosynthesis. The flow of carbon atoms is toward the wing tips. In hard waters, calcium carbonate often precipitates simultaneously with removal of carbon dioxide. The "wings up" position corresponds to the opposite situation— the addition of carbon dioxide to water as a result of the cooling of water in autumn or due to the respiratory activity of living organisms at night. The flow of carbon atoms is from the wing tips to the body and bicarbonate increases at the expense of carbon dioxide and carbonate ions. If calcium carbonate is present it tends to dissolve, although only slowly. Precipitates usually dissolve much more slowly than they form.

The mechanical bird can now be put to good use in explaining what would otherwise be an unusual aspect of human-caused eutrophication— why soft-water lakes respond to nutrient enrichment with dramatic increases in pH, whereas calcium carbonate precipitates in hard-water lakes with less change in pH.

The first response to nutrient enrichment in natural waters is usually an increase in the rate of photosynthesis and abundance of plants, including algae. Exceptions can occur, for example, when waters are so turbid that light penetration limits the depth of photosynthesis or water flows through the system so rapidly that nutrients and photosynthetic products are

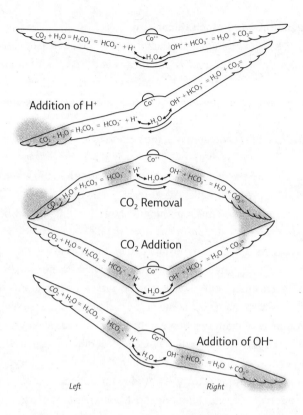

FIGURE 4.1: *A mechanical bird, showing the relationship between various forms of inorganic carbon in water under various conditions. Dots indicate the approximate relative abundance of different forms of carbon, with molecules flowing "downhill," depending on the wing angle with respect to the vertical. Removal of dots (molecules) from the wingtips corresponds to passage of $CO_2$ from water to air (left wing) or formation of calcium carbonate (right wing). Arrows under the body of the bird indicate the tendency of water ($H_2O$) to generate hydrogen ions ($H^+$) or hydroxyl ions ($OH^-$).* From Vallentyne (1974).

washed rapidly out of the systems. In hard waters, increased photosynthesis leads to a precipitation of calcium carbonate because of the uptake of carbon dioxide by photosynthetic plants, as already described. In soft waters, the concentration of bicarbonate is so low that it cannot act effectively as a buffer for hydroxyl ions released by the photosynthetic uptake of carbon dioxide. As a result, hydroxyl ions accumulate, accounting for the dramatic rise in pH.

These responses are not just theoretical predictions from armchair scientists. They have been thoroughly analyzed and described in a number of lakes. One of the more interesting cases on record is that of Lake 227, one of 46 small lakes[2] set aside for experimental studies on water pollution. Lake 227 is a soft-water lake. As a result of the increased abundance and photosynthesis of plants, brought about by controlled additions of phosphate and nitrate salts over a period of 2 years, the pH rose from 6 to 10, representing a 10,000-fold increase in the concentration of hydroxyl ions. A more detailed treatment of chemical changes accompanying the rise in pH is given in Chapter 9. The precipitation of calcium carbonate as a result of human-caused eutrophication in a hard-water lake has already been described for Lake Zürich in Switzerland. Precipitation of calcium carbonate there was merely one of a sequence of well-documented events, later found to be characteristic of nutrient-enriched lakes all over the world.

The equation chemists use to describe the mutual reciprocity treaty that nature has imposed upon the realms of photosynthesis and respiration shows water and carbon dioxide on one side of the equation and organic matter and oxygen on the other. Water and carbon dioxide have already been dealt with. It remains now to consider organic matter and oxygen.

## Life in Inland Waters

JUST AS VARIOUS TYPES of natural plant communities develop on land in response to local conditions of climate, topography, and soil, so are there natural communities of aquatic organisms that flourish under particular conditions within lakes. With two major exceptions, the important physiological factors in the aquatic environment are the same as those on land—light, temperature, water currents (instead of wind), and the chemical and textural nature of the environment. The two major differences are water, which is omnipresent in lakes but in variable abundance on land, and oxygen, which is constant in the air above land but in variable concentrations in the waters of lakes. In surface waters, photosynthesis during daylight can produce oxygen faster than it is utilized by respiration and released at the lake surface, causing supersaturation of the epilimnion with oxygen. At night in productive waters, there is no photosynthesis because of the low light. As a result, oxygen can decrease to below saturation as the result of respiration and decomposition. In extremely productive waters where respiration is high, oxygen can decrease to zero overnight. The amount of oxygen in water at saturation also varies, with cold water containing more oxygen than warm water. At near-freezing temperatures, water saturated with oxygen contains over 14 ppm of $O_2$, and at 25°C, saturation is only a little over 8 ppm. Because respiration and decomposition are also higher at warmer temperatures, oxygen can be depleted overnight in productive waters during summer, to the detriment of air-breathing organisms. In the deep waters of lakes, there is generally a cumulative seasonal depletion of oxygen caused by decomposition of sinking plankton and bacterial respiration, which can range from a few tenths of a part per million in oligotrophic lakes to complete anoxia in highly eutrophic waters. We will discuss the consequences of anoxia for nutrient recycling in Chapters 9 and 10.

Three primary lifestyles have been adopted by organisms in aquatic ecosystems, two associated with open-water conditions and one with bottom deposits. *Plankton is* the community of organisms with limited powers of locomotion inhabiting the open-water environment and mostly living independently of the bottom. The distribution of planktonic species is in large part regulated by water currents. Plants in the plankton are generally species of algae, collectively referred to as *phytoplankton*. The animal equivalents are mostly small crustaceans and rotifers, collectively called *zooplankton*. The other group of organisms, inhabiting open-water regions of lakes, consists of fish and other vertebrates that move about "under their own steam." Scientists refer to these as *nekton*, although the term is rarely used by non-specialists. Organisms living in association with bottom deposits are collectively referred to as *benthos*. As for the plankton, they can be separated into *phytobenthos* and *zoobenthos* for plant and animal forms, respectively.

Most planktonic species are smaller in size than an average grain of beach sand, but some (such as the larva of the phantom midge, *Chaoborus;* the crustacean *Mysis relicta;* and the widely distributed but rarely seen freshwater jellyfish, *Craspedacusta;* the predatory cladoceran *Leptodora kindti;* and the alien invading crustacean *Bythotrephes*) reach 1 centimetre or more in length. Many benthic organisms are larger on the average than planktonic forms; they consist of algae that grow attached to rocks, logs, and other substrates near the shore (termed *periphyton*), pondweeds (generally called *aquatic macrophytes*), and a variety of aquatic insects and other invertebrate animals.

In the daily drama of life in inland waters, planktonic forms move restlessly about in the water in search of food or a mate, using their primitive instincts to avoid being trapped in unfavourable environments. Prey species must be constantly on the lookout for predators ready to pounce on an unsuspecting individual at a moment's notice. Death is present as a fact of life. In nature, many species lay hundreds or thousands of fertilized eggs, even though an average of only two will survive.

The total number of species of plants and animals in freshwater ecosystems is greater than one might initially think. It ranges well into the thousands for any single lake and into the tens of thousands for inland waters as a whole. New species are still being discovered and described at a rate of several hundred per year. The detailed life histories and interactions of these forms are so complex that an encyclopedia could not fully encompass even the little that is known. In the complexity of this pattern, there lies a common thread that serves to simplify and unite. This is the story of the eater and the eaten, in which similarities of feeding relationships permit organisms to be grouped into a limited number of *trophic* or feeding levels in a food chain.

## Trophic Levels

THE UTILITY of this *trophic* approach to ecology is that it permits a dynamic understanding of how energy from the sun flows through successive levels of the food chain—from the sun to green plants, from green plants to herbivores, and from herbivores to carnivores and decomposers. It also serves as a reminder of how organisms in different trophic levels depend on each other. In short, it permits the economics of ecosystems to be assessed.

The trophic level at the base of the food chain consists of green plants, including algae. They use the sun's energy to synthesize organic compounds, used as food by organisms at the next trophic level. Expressed on an annual basis, the net efficiency of this energy conversion is rather low in most freshwater ecosystems. Firstly, only about 50% of solar energy can be used in photosynthesis. UV and IR wavelengths are not used, and even in the visible range, some wavelengths are more efficiently used than others. Photosynthesis generally ranges from 0.01% to 1.0% when expressed as calories of radiant energy converted into organic material by green plants in the course of a year. This means that 99.0-99.99% of the sun's energy is not utilized in photosynthesis. Further reasons for these low efficiencies are that (*i*) light is absorbed by nonliving things (water, suspended clays, dissolved colouring matter) before it reaches the cells of green plants, and that (*ii*) nutrient deficiencies can limit photosynthesis when light is abundant.

Fortunately for fish, in recent years, scientists have found that much of the production of higher trophic levels is fuelled by allochthonous carbon. As described above, the catchments of lakes supply dissolved organic matter. They also supply organic carbon in particulate form as leaves, needles, stems, and small organic particles that wash or blow into lakes. Generally, many of these particulate sources settle in shallow water, where they can be eaten by benthic invertebrates, or bacteria. It has been found that the different mechanisms involved in metabolism of terrestrial plants leaves a distinct stable carbon signal in the carbon of lakes that can be used to separate the pathways for allochthonous and autochthonous carbon sources. In small- to medium-size oligotrophic lakes, the allochthonous carbon sources can contribute as much or more to the energy required by fish as the autochthonous production. Recent studies employing stable isotopes of carbon to "fingerprint" where the energy in aquatic organisms originates. It is beyond the scope of what we discuss here, but interested readers can find examples in Hecky and Hesslein (1995) and Pace et al. (2004).

Although the low efficiencies of transfer of energy in aquatic food chains may be unwelcome news to anglers who value lakes for fish, the increased plant abundance would be noted with distaste in every sip of water or swim at the beach if aquatic ecosystems were more efficient photosynthetically than they are. Sometimes even fishermen's interests can be adversely affected—as in the case of lake trout (*Salvelinus namaycush*), where an increased plant abundance can mean more fish but less trout, because the plants eventually die and sink into the cold hypolimnion where trout spend their summers. As the plants are decomposed, the oxygen in this zone becomes depleted.

The trophic level at the next position in the food chain consists of herbivores—animals that feed on green plants. In the planktonic community, these consist of small forms that strain algae through hairy appendages specifically designed for filter feeding. To appreciate the fantastic anatomical design that evolution has created one should travel with a modern Gulliver into the miniature world of nature as seen through a microscope. Benthic herbivores feed by scraping algae from the surfaces of rocks and pondweeds or by munching on leafy vegetation. They also feed on detritus, minute remains of dead plant and animal tissues.

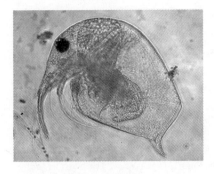

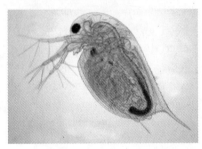

FIGURE 4.2: *Genera of various crustaceans that are common in the plankton of lakes. Top: the cladoceran* Bosmina. *Length 0.5–0.8 mm. Middle: the cladoceran* Daphnia. *Length 1.5–3 mm. Bottom: the copepod* Hesperodiaptomus. *Length 2–3 mm.*
Photographs courtesy of Mark Graham (top), Howard Webb (middle), and DWS (bottom).

The net annual efficiency for conversion of plant material into herbivore growth varies, but is typically less than 10% in lakes, sometimes far less. This efficiency is much higher than for conversion of the sun's energy into new plant growth. This is because the plant energy that herbivores consume is mostly in a readily assimilable chemical form.

Typical herbivores in the plankton of a lake include Cladocera (Crustacea), Copepoda (also Crustacea), and rotifers (Rotifera). Figure 4.2 shows examples of two cladocerans and a copepod. With few exceptions, Cladocera and Rotifera are herbivores. The Copepoda have larval stages called nauplius larvae that are also herbivorous. The more advanced copepodid larvae and adults of some copepods are also herbivorous (although they may eat protozoans and small crustaceans as well), but some larger species eat Rotifera, small Cladocera, and nauplius larvae as adults.

Zooplanktivores, animals that prey on herbivores, occupy the next level of the trophic chain. Many predacious insects, crustaceans, minnows, and other small fish occupy this trophic level. Fish tend to focus on larger members of the plankton, such as large, slow-swimming *Daphnia* or large Copepoda that can be brightly coloured and easy to see.

In turn, zooplanktivores are eaten by higher carnivores, including piscivorous species such as lake trout, bass (*Micropterus*), walleye (*Sander vitreus*), and pike (*Esox lucius*). As the mouths of carnivores get larger, carnivores that prey on carnivores come into play, exemplified by pike and lake trout. The net annual conversions of herbivore energy into planktivore and planktivore to piscivore growth lies in the same range as conversion of plant energy into herbivore growth—about 10% at each step in many lakes.

As a result of high energy losses associated with the passage of energy through each level of the food chain, the amount of plant material required to produce a given weight of carnivorous fish can be immense. In order to add one unit of weight to a lake trout, for example, the energy equivalent to several hundred comparable units of weight of plant material has to pass through the food chain. The situation is similar in the food chain from grass to cattle, via beefsteak to humans.

## Decomposition

AS BRIEFLY MENTIONED ABOVE, organic matter produced by aquatic or terrestrial plants and not eaten by herbivores is acted on at other levels of the food chain by bacteria, fungi, and protozoans. These microorganisms perform essential ecosystem services, somewhat like the role a vacuum cleaner plays in respect to dust and dirt in an office or house. Many of these microorganisms transform tiny particles and dissolved organic molecules into particles that again become large enough to enter the food chain leading to higher animals, via the trophic pathway known as the "microbial loop." In the normal course of their daily lives, most people are unaware of the existence of these microorganisms, but they soon would be if the microbes ceased to function. The decomposable wastes we throw out as garbage or flush down the drain would accumulate in the natural environment to a most unpleasant degree.

When people "size up" material objects, as the saying goes, they tend to attribute importance of an object in proportion to how big it is, without ever thinking too much about what it does. It would surprise most to learn that minute, and sometimes almost invisible, populations of algae in eutrophic lakes can produce as much or more organic matter in a year than trees in an equivalent area of neighbouring forest. They accomplish this by "turning over" their populations at a rapid rate.

As a general rule, the smaller the organism or growth stage, the higher the rate of growth is per unit weight. Thus, under favourable conditions, bacteria can reproduce their original weight in a matter of minutes or hours; unicellular algae, in hours or days; but an insect larva takes days to weeks and a large fish or rooted plant, months to years. The same is true of respiratory processes. As one passes from the small to the large in the living world, the rate of metabolism per unit weight declines logarithmically as the weight of the individual increases. For these reasons, one can look to microscopic species as the predominant organisms regulating the "breathing" of lakes. Algae are commonly the most important primary producers of organic matter and oxygen. Bacteria, fungi, and other small organisms are the most important agents of decomposition, consuming oxygen in the breakdown of organic matter and producing carbon dioxide. The minuteness of their size is deceiving in terms of the important ecosystem functions they perform.

The depth range over which photosynthetic oxygen production occurs in lakes depends on the transparency of the water. It can be as little as a few centimetres (inches) in very turbid waters, to as much as 30 metres (100 feet) or more in extremely clear lakes such as Lake Baikal in Russia; Crater Lake, Oregon; and Great Bear Lake in northern Canada. However, in most lakes, the bulk of all photosynthesis occurs in the epilimnion. The hypolimnion is a region of net decomposition.[3]

## Hypolimnetic Oxygen Depletion

BECAUSE THE BOUNDARY between the zones of net photosynthesis and decomposition frequently coincides with the thermocline, one can readily appreciate the problem that commonly arises. As organisms die in the epilimnion, they slowly sink through the metalimnion into the hypolimnion, where they decay. Cut off from contact with the atmosphere by the barrier of the thermocline and with too little light for active photosynthesis, the hypolimnion is a zone where oxygen is consumed but cannot effectively be regenerated. As a result, the supply of dissolved oxygen tends to become progressively depleted in the hypolimnion with time.

The extent of oxygen depletion in the hypolimnion varies with a number of factors. The three most important of these are (i) the rate at which organic material falls into the hypolimnion (which is affected by the rate of photosynthesis, in turn usually determined by the supplies of essential nutrients like phosphorus and nitrogen), (ii) the volume of the hypolimnion relative to the dimensions and shape of the lake basin, and (iii) organic matter yield from the lake's catchment. The first factor is primarily determined by the rate of photosynthetic production of organic matter in the epilimnion; the second, by the extent to which the lake basin acts as a funnel in concentrating the "rain" of dead and dying plankton from above; and the third, by the size, productivity, and slope of a lake's catchment. The situation is comparable to making a pot of tea. The more tea leaves (that is, dead or dying phytoplankton or terrestrial organic matter) and the smaller the volume of water used in making the tea (that is, volume of the hypolimnion), the darker the final product (that is, the higher the concentration of organic matter and the greater the consumption of oxygen in the hypolimnion). In general, in very large and deep lakes, oxygen depletion in bottom waters is not extensive because of the large volume of the hypolimnion and large initial supply of oxygen.

Because the consumption of oxygen in decomposition is matched by respiration, in general every mole of oxygen consumed is matched by a mole of $CO_2$ produced. Thus, as oxygen is depleted in the hypolimnion, $CO_2$ accumulates (Figure 4.3).

The extent of oxygen depletion and carbon dioxide formation in bottom waters of eutrophic lakes intensifies with the progression of summer. Because bacteria and organic matter are concentrated at the sediment surface, oxygen depletion and $CO_2$ formation typically proceed "from the bottom up." When the concentration of dissolved oxygen falls to less than 1 parts per million just above the sediment surface, new chemical reactions that are relevant to eutrophication comes into play, first elaborated by Wilhelm Einsele in Austria and Clifford Mortimer in England. Insoluble iron hydroxide in the sediment changes from a ferric ($Fe^{3+}$) to a ferrous ($Fe^{2+}$) form, inducing a change of sediment colour from brown to black.[4] At the same time, various chemicals are released from the sediment into the water, moving upward and laterally as a result of turbulence induced by internal seiches. Under such conditions, organisms that depend entirely on oxygen disappear from the sediment surface by migration, encystment, or death. Microorganisms that can adapt to low oxygen, including new

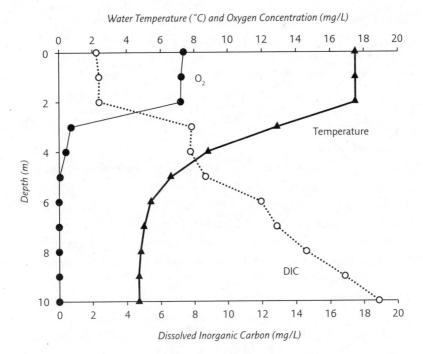

*Water Temperature (°C) and Oxygen Concentration (mg/L)*

FIGURE 4.3: *Typical profiles of temperature, oxygen and dissolved inorganic carbon (DIC), the sum of $CO_2$ and bicarbonate ion in a lake during late summer stratification (Lake 227, Experimental Lakes Area, Ontario, Canada).*

species, replace anaerobes. In general, bacteria that survive at low oxygen can metabolize by removing oxygen from other chemicals like $CO_2$, sulfate, or nitrate. If the zone of oxygen depletion extends upward to the thermocline, a heavy mortality of deepwater fish can occur.

With the breakdown of the thermocline in autumn, a reverse process takes place. As the waters of the lake are mixed and re-oxygenated, iron hydroxide precipitates. Most other constituents released from sediments decline in concentration, through dilution induced by mixing with surface waters in some cases and by precipitation in others. Some nutrients liberated from the sediment remain in the water to provide nourishment for later algal growth. The importance of nutrient regeneration from sediments in promoting plant growth varies greatly, depending on a lakes depth, productivity, and degree of thermal stratification. In some lakes, it is very important and of little consequence in others.

As described in Chapter 9, the additional respiration in lakes fueled by the use of allochthonous organic matter also produces $CO_2$. As a result, lakes generally emit more $CO_2$ per year than can be accounted for by autochthonous production alone. This supersaturation with carbon dioxide is a common feature of lakes (Schindler et al. 1975; Cole et al. 1994; Kelly et al. 2001).

## Winterkill Lakes

MOST CHANGES DESCRIBED ABOVE for the bottom waters of eutrophic lakes in summer also occur under the cover of ice and snow in temperate lakes. In shallow, eutrophic waters, this can lead to a "winterkill" of fish because of oxygen depletion in the entire body of water. An account of the culture of rainbow trout in "winterkill" lakes will illustrate how lakes "caught short of breath" due to lack of oxygen can be utilized and developed by humans.

In the prairies of western Canada, there are numerous shallow, naturally productive ponds and lakes. Because of their limited depth, rich nutrient supply, and high rates of biological production, they typically become depleted in dissolved oxygen under ice and snow. They are known as "winterkill" lakes because concentrations of oxygen at the end of the long prairie winters are frequently so low fish do not survive. Fishery workers naturally avoid such lakes when planting young hatchery-reared fish.

The absence of predatory fish in winterkill lakes and the favourable growing conditions of prairie summers permit some bottom-living invertebrates to become unusually large and very abundant. This is particularly true of *Gammarus*, a shrimp-like crustacean that can survive the low oxygen conditions of winter. The waters teem with algae because of the high nutrient supply, and are rich in large invertebrates because of the abundant supply of plant food and lack of predators.

With the genius it takes to recognize the obvious, a group of Canadian scientists decided to see if winterkill lakes could be put to use in rearing marketable fish. The logic was simple—introduce fast-growing fish as fingerlings in the spring and harvest them at marketable size in the autumn. The species selected was rainbow trout (*Oncorhynchus mykiss*). Some trout were planted as 5- to 7-centimetre (2- to 3-inch) fingerlings in the spring of 1968 and harvested as 22- to 35-centimetre (9- to 14-inch) fish in the autumn. They ranged in weight from 0.3 to 0.45 kilogram

(0.75–1 pound). Thus, the critical winter period when oxygen depletion under ice would have killed fish was avoided.

The prairie fishpond industry has remained small, because it is limited by the size and number of suitable ponds. Some ponds are too productive, so that respiration at night uses all the oxygen, and fish do not survive. In other cases, fish grow well, but have a peculiar taste that makes them unmarketable. However, in suitable ponds, yields of fish per unit area are comparable with yields of prairie wheat in nearby fields. With genetic improvements in fish brood stock comparable with those for hybrid corn, it was once hoped that the Canadian prairies might one day become famous for naturally reared rainbow trout. However, prairie trout culture has remained small.

This terminates the discussion on regulation of carbon dioxide, oxygen, and organic matter in aquatic ecosystems and also the diversion into limnology that began with an account of the importance of water in lakes. The story turns next to phosphorus in the biosphere, as a prelude to a systematic analysis of how cultural eutrophication can be brought under control.

# 5 | Phosphorus, The Morning Star

*The role of phosphorus in the economies of nature, life, and humans*

LIKE MOST ALCHEMISTS of the 17th century, Hennig Brand of Hamburg was possessed with a desire to discover the secret principle of the philosopher's stone, the mythical material reputed to have the power to transform base metals into gold. Although unsuccessful in this quest, Brand brewed a concoction in 1669 that yielded a no less mysterious product: one that, unlike any other substance known at the time, glowed in the dark by its own power and ignited spontaneously when exposed to air. He called it phosphorus, meaning light-bearing, from the name given by ancient Greek and Roman astronomers to the planet Venus when it appeared in the sky as the morning star.

One can easily imagine the stir Brand's discovery must have created at the time, kindling each alchemist's hidden belief that surely, with such a momentous finding as this, the discovery of the philosopher's stone and the chemical elixir of life could not be far away. However, chemistry at the time was more of an art than a science. None of its modern foundations had been laid. One hundred years would elapse before the nature of water would be understood as a compound of hydrogen and oxygen, rather than the element for which it was then mistaken. A century and a half would pass before the first chemical synthesis of a biological compound. In 1669, the periodic table of chemical elements had yet to be conceived, the structure of atoms and molecules remained unknown, and acids and bases were described as male and female. Even air was looked upon as a single substance, as in Aristotle's time 2000 years earlier.

It is significant to cultural eutrophication that Brand used urine as the starting material for his synthesis of elemental phosphorus. By evaporating a large volume of urine to dryness and heating the residue in the presence of fine white sand, he watched mysterious vapours rise into a second vessel where yellow-white crystals of elemental phosphorus formed under a protective layer of water. Little did Brand know that in his failure to resolve the riddle of the philosopher's stone, he had accidentally discovered a chemical element that 300 years later would be regarded as one of the keys to the regulation of plant growth in nature. Viewing the situation now from the vantage point of later time, let us turn back the clock to retrace some of the events that led slowly, yet inevitably, to the recognition of the central role of phosphorus in life.

After the initial burst of activity that followed Brand's discovery of elemental phosphorus, progress in unravelling the biochemistry of its compounds was slow. The first demonstration of the presence of phosphorus in organisms was made in 1771, by the Swedish chemist, Karl Wilhelm Scheele, who identified it as a major constituent of bone.

The next important event was in 1840, when Justus von Liebig, the founder of organic chemistry, showed that the fertilizing effect of humus on plant growth was due to inorganic salts of phosphorus and nitrogen rather than to organic matter, as previously assumed. This discovery opened the path for studies on soil chemistry in relation to plant nutrition that eventually led to modern methods of crop production.

### The Historical Setting

THE 19TH CENTURY was a time of deep intellectual ferment in regard to the nature and significance of life. Toward the close of the previous century, a momentous discovery had been made by Antoine Lavoisier, father of modern chemistry, whose unjust reward for a life of service to France was the loss of his head on a guillotine.[1]

Lavoisier and Pierre Laplace showed that the amount of heat produced by animals from the consumption of a given amount of food was similar to that produced by chemical oxidation of the same weight of material in the laboratory. This demonstration was of such monumental importance that over 150 years were required for full documentation of the intermediary biochemical events.

Isolation and characterization of organic compounds from living organisms was also proceeding slowly, yet persistently, in the 19th century. Interest accelerated following the synthesis of urea by Friedrich Wöhler in 1828. This was the first widely recognized chemical synthesis of a biological compound. Profound biological generalizations and discoveries were being made. Cells were recognized as the fundamental units of life. New species and morphological structures in plants and animals were being categorized systematically. It was gradually becoming apparent that all the architectural complexity of life had a mechanistic basis.

Foundations for this belief were greatly strengthened with the publication of Darwin's *The Origin of Species* in 1859. Darwin drew attention to the immense changes created by humans in domesticated species of plants and animals. If human selection could produce such pronounced differences over a relatively short time, as seen in the various breeds of horses, pigeons, dogs, cats, and cultivated plants, how much more infinite would be the variety that natural selection could produce over longer periods of geologic time?

With the appearance of the *The Origin of Species,* those who previously accepted humans as unique and divine in origin suddenly found themselves in the embarrassing position of having to re-evaluate their interpretations of humans and life. If humans had evolved from an apelike form, as was implicit in *The Origin of Species,* then what was the significance of humans? In 1632, Galileo tormented the conceit of humans with his proposition that the Earth was not the centre of the solar system. Giordano Bruno was burned at the stake by the Catholic Church in 1608 for this and the even more far-reaching belief that life might not be peculiar to Earth. Then, in the middle of the 19th century, Darwin had thrown the very soul of humans onto a bed of evolutionary thorns by proposing that *Homo sapiens* originated by purely natural laws from a continuous line that led back to the lowest forms of life.

*The Origin of Species* revived an age-old scientific discussion about the significance of life. Vitalists claimed special forces were present in living cells that could not be explained in terms of the laws of physics and chemistry. Mechanists, on the other hand, were inclined to regard living organisms as highly complex self-regulating machines, whose inner workings would one day be explained in purely mechanical terms.

The discovery that provided the factual basis for the resolution of these widely divergent claims was made by Edouard Buchner in 1897. Buchner obtained a cell-free extract of yeast, capable of carrying out the process of alcoholic fermentation. With this demonstration of enzymes (specialized proteins that accelerate chemical reactions in living cells), the science of biochemistry arose. In the years following Buchner's discovery, enzymes isolated from various tissues and organisms were studied to determine the mechanisms of catalytic action. As knowledge developed, it became clear that similar enzymes and biochemical cycles were present in organisms as diverse as yeasts and humans. With this came the recognition that life had certain common and fundamental biochemical characteristics, on which the immense anatomical diversity was based.

## The Many Roles of Phosphorus in Living Organisms

As these pieces successively began to fit into place, it became more and more evident with each new discovery that compounds of phosphorus played unique and essential roles in all the fundamental biochemical reactions of life—photosynthesis, respiration, fermentation, muscular contraction, cell division, and heredity. Eventually, phosphorus was recognized along with hydrogen, oxygen, carbon, nitrogen, and sulfur as one of six principal elements that form the backbone of life.

Of several hundred phosphorus compounds isolated from living cells, three deserve particular mention. Two of these, DNA (deoxyribonucleic acid) and RNA (ribonucleic acid), are actually not compounds but general classes of polymers. The third is a small molecule, ATP (adenosine triphosphate), that functions as a fuel that drives energy-requiring synthetic reactions in cells.

DNA was first discovered in 1869 by Johann Friedrich Miescher, who noticed a weakly acidic substance of unknown function or structure in the nuclei of human white cells. He called the substance deoxyribonucleic acid, or DNA. The tool that was ultimately to reveal the role of DNA in reproduction was developed many years later, after the discovery and application of X-rays. In 1912, Sir William Henry Bragg and his son Sir William Lawrence Bragg discovered that the structure of crystals could be deduced from their diffraction patterns when bombarded with X-rays. X-ray diffraction is the tool that ultimately allowed the structure of DNA to be discovered, but two other important steps occurred first. In 1928, Franklin Griffith, a British medical officer, discovered that genetic material from heat-killed bacteria could be transferred to living ones. He called the process transformation. In 1949, Erwin Chargaff, a biochemist, discovered that DNA was different between different species. The base adenine was always found to be equal to thymine, and the base guanine was equal to cytosine, but the amount and proportion of these base pairs differed among different species. Many remained skeptical that a molecule with such simple composition could be the basis for reproduction in the many complicated forms of life. Proof was provided in 1953 by James Watson and Francis Crick, who showed how the simple properties deduced by Chargaff could contain enough information to allow differences between species and individuals, via the structure of the famous "double helix" structure of the DNA molecule. It is now recognized that the extremely high quality X-ray diffraction photographs of DNA taken by British crystallographer Rosalind Franklin were instrumental in Watson and Crick's discovery.

DNA is located in chromosomes within the nuclei of cells, where it constitutes the ultimate units of heredity, genes. Although both DNA and genes were under study in the latter part of the 19th century, it was only in the 1950s that the identity of genes as DNA molecules was firmly established.

DNA is perhaps the most remarkable of all compounds humans have yet discovered. Without it, no cell, individual, or species could reproduce its internal complexity. Not even Johann Sebastian Bach in a million lifetimes could invent as many variations for the keyboard as natural selection has already produced in the DNA of the chromosomes in living species. One set of those delicate filaments, enclosed within a space a million times more microscopic than a grain of sand, stores the complete information necessary to create an adult organism from a single cell. When the thread of DNA for a given species is terminated by the death of the last breeding pair—as has happened to the passenger pigeon, Tasmanian wolf, and several hundred other species that have vanished during the past 200 years—nothing can recall into existence what was there before.

In contrast to DNA, RNA occurs mostly in the cytoplasmic (non-nuclear) parts of cells. Several different kinds of RNA molecules occur in living cells. Some transfer the chemically coded information from DNA to sites of protein synthesis. Others are involved in specifying and assisting in construction of cellular proteins. Different types of RNA molecules operate like managers in a factory, controlling the location and manufacture of parts in such a way that new cells are produced. From knowledge contained in the chromosomal books of instruction that is peculiar to each individual and species, they supervise and control the synthesis and operation of enzymatic machines.

To enumerate other phosphorus compounds in living cells without reference to ATP (adenosine triphosphate) would be like listing all of the forms of energy that can be converted into electricity. Long before human societies evolved, organisms discovered the utility of utilizing one form of energy to run a variety of machines. In cells, this energy is chemical. It is called ATP.

When the terminal phosphate group is removed from ATP, energy is released that can be used for muscular contraction, the synthesis of energy-rich compounds, or growth. In the process, ATP is converted to ADP (adenosine diphosphate) and a phosphate ion. The circle is complete when various carbohydrates, fats, and proteins are broken down to carbon dioxide and water in cellular respiration. Energy released in the breakdown is used to resynthesize ATP from ADP and phosphate. In this way, countless different organic fuels are converted into a single form that serves various energy-requiring processes in cells. As is the case in most biological systems, reactions of ATP and ADP are cyclic—the products of one reaction serve as the starting material for another that, in turn, regenerates the first.

## Phosphorus in the Biosphere

The cycle of phosphorus in the biosphere is primarily regulated by three factors: biological uptake, precipitation reactions with certain positively charged ions, and a net flow to sediments and the sea. In most natural occurrences, inorganic phosphorus atoms are fully complexed with oxygen atoms to form phosphate ions.[2]

Phosphoric acid ($H_3PO_4$) dissociates into hydrogen ions and several different kinds of phosphate ions to an extent that depends on concentration and the pH of the medium. Since all the different phosphate ions are interconvertible depending on pH, there is no need to bother with their names here.

In general, phosphate compounds tend to be soluble in water except in the case of minerals formed by interaction with calcium, iron, aluminum, and some other ions. These react with phosphate ions at neutral or high pH to form precipitates that have a low solubility in water. The reactions are as follows:

calcium ions + phosphate ions ⇌ calcium phosphate ↓

ferric ions + phosphate ions ⇌ ferric phosphate ↓

aluminum ions + phosphate ions ⇌ aluminum phosphate ↓

The first reaction predominates in calcareous soils (soils that are rich in calcium and carbonate). The reaction is more complicated than one might first think, because it leads to the formation of apatite minerals, complex crystals that also contain other ions.[3] In lime-deficient soils, phosphate ions principally react with iron and aluminum ions to form salts of low solubility. In summary, the chemical behaviour of phosphorus is governed by geochemically abundant ions, principally those of calcium, iron, and aluminum, that react with phosphate ions to form precipitates.

Living organisms also play a major role in regulating concentrations of phosphorus in nature. Phosphates are rapidly metabolized in the soil by microorganisms and are taken up with other nutrients by the "root hair" cells of terrestrial plants. Along with nitrogen and potassium, phosphorus is one of the elements most commonly depleted in cultivated soils by the removal of crops. This is why it is a major component of fertilizers.

In the surface water of lakes in temperate latitudes, dissolved phosphate tends to pass through seasonal cycles of concentration. These cycles are largely controlled by seasonal changes in sunlight and temperature and by activities of green plants. Dissolved phosphate accumulates in winter when photosynthetic activity is low, and decomposition releases it from organic matter. Later, in spring and summer, dissolved phosphate in the epilimnion typically becomes almost undetectable because of its removal by photosynthetic plants.

Sometimes, people not familiar with lakes are surprised to hear that the concentration of dissolved phosphate is often low prior to the appearance of algal blooms. Their first thought is that phosphorus must be present in an adequate supply, because the blooms occur in spite of the low concentrations. What they forget, or perhaps do not know, is that Cyanobacteria (called blue-green algae, although technically they are not algae because they lack chloroplasts) can store phosphate in excess of their immediate needs, utilizing the excess under conditions of phosphorus deficiency at a later time. Like the empty dish after a good meal, the concentration of dissolved phosphate in water is low just prior to the appearance of algal blooms because the phosphate is in the algae.

The rapid rate at which dissolved phosphate ions are removed from surface waters by plants is indicative of the active role they play in lakes. When radioactive phosphate is added to the surface waters of lakes, it is taken up by algae within a matter of minutes. Later, the death and sedimentation of algae results in a slow transport of phosphorus from water to sediments. In contrast to the rapid uptake of dissolved phosphate ions from water by plants, a few weeks or months may be required for sedimentation.

The overall cycle of phosphorus in nature is essentially a one-way drainage to the sea. No volatile compounds containing the element exist that can be returned to land through atmospheric transport and precipitation as is the case for carbon and nitrogen. The little recycling that does occur is brought about by the slow geologic upheaval of marine sediments and, on a shorter time scale, by the removal of fish from the sea by humans and fish-eating birds, by the migration of salmon back to freshwater where they die, or by the mining of phosphate-rich marine deposits by humans for fertilizer.

The world's supply of high-grade phosphate rocks is currently estimated to be about 50 billion tons, mostly in the form of apatites—complex calcium phosphate minerals that also contain hydroxyl and fluoride ions. There are three principal areas of supply: Morocco and adjacent regions, an area west of the Ural Mountains, and northwestern and southeastern parts of the United States. Some of these deposits were formed by igneous processes that took place deep within the Earth. In other cases, they consist of marine sedimentary rocks. No counterpart to processes that, in former times, caused sedimentary phosphate deposits has been discovered in modern oceans.

The world's reserves of phosphate rocks are sufficient to fulfill current demands for what some estimate as hundreds, and others an indeterminate number, of years to come. The significance of this time, whatever it may be, hinges on technology and how one views a nonrenewable resource in terms of time.

Phosphates enter into the economy of humans in many different ways. Their principal uses in the United States in 1967 were 70% in fertilizers; 13% in detergents; 8% in animal feeds; and the remaining 9% distributed among a variety of applications such as leavening agents in baking powders and cake mixes, anticorrosion agents for car bodies, fireproofing materials, chemical agents in some types of water softeners, smoke bombs and incendiaries, insecticides, rodent poisons and lethal nerve gases, additives to gasoline, "muds" used in oil drilling operations, bottled "pops" and tonics, toothpastes, dental cements, matches, and plasticizers. Note how the uses range from life-promoting fertilizers to deadly poisons intended for rodents, insects, and humans. One drop of sarin, a lethal nerve gas containing phosphorus, if not immediately washed from the skin, can kill a human as rapidly as a bullet through the brain.

## The Toxic Properties of Elemental Phosphorus

IN CLOSING THIS DISCUSSION of phosphorus in the economies of life, nature, and humans, the hands of the clock turn back to where the account began—to Hennig Brand's separation of elemental phosphorus in 1669. A modern industrial process, based on the reaction he discovered, gave rise in 1969 to a horror story of industrial pollution—one that contrasts the design of nature and the magic of life with the behavioural attributes of humans. Much of this industrially produced elemental phosphorus was destined for use in detergents. The following is summarized from Jangaard (1970), Hodder et al. (1972), and Chamut et al. (1972).

The story begins in the 1960s in Long Harbour on the eastern shore of Placentia Bay, Newfoundland, where a Canadian company established an industrial plant to produce elemental (yellow) phosphorus from phosphatic rocks imported from Florida. The company located the plant in Newfoundland because of the guarantee of cheap electric power from a province wanting to diversify its economic base.

Much to their dismay, the fishermen of Placentia Bay awoke on the morning of December 12, 1968, to find marine herring (*Clupea harengus*) floating belly-upward in Long Harbour with extensive hemorrhaging of blood vessels about the gills and fins. Because of this redness about the head and fins, the fish became known as "red herring." Fishery inspectors were notified; however because they hadn't encountered a case of "red herring" before, they didn't quite know what to do, except to explore the area in the vicinity of the plant. Samples were frozen and sent to biological experts for study, but even the experts were puzzled. None of the usual factors involved in fish mortality seemed to be present.

The water in which the dying fish were found (at Long Harbour, adjacent to the industrial plant) was saturated with oxygen and did not contain excessive amounts of organic wastes. Pathogens and parasites were lacking in the fish. The only symptoms were the hemorrhaging about the gills and fins, and the watery nature of the blood in dying fish. Although nothing like it had been heard of before, one significant fact was established by divers on January 27, 1969. The entire bottom of Long Harbour, to a depth of 4.5 metres (15 feet), was devoid of animal life. On the suspicion that something very serious was wrong, the area was closed to fishing, and fishermen advised to destroy any catches that contained "red herring."

No further reports of "red herring" occurred until February 1969, when they were observed off the mouth of Long Harbour. Then, with surprising rapidity, occurrences started to spread to distant locations. All up and down the coast, large numbers of "red herring" were seen wandering blindly in surface waters. Fishermen were afraid to eat their own fish, and rumors began to circulate of poisonous gases said to have been dumped into the bay after the close of World War II. Though nothing was certain, public suspicion was directed at the industrial plant.

Suspicions of scientists involved in the investigations were also directed at the plant. The number of coincidences was too high to be accidental. In the first place, the mortality on December 12, 1968, included a variety of animal life in addition to herring. Secondly, it occurred in Long Harbour in direct proximity to the industrial plant. Thirdly, the date December 12 (as it later turned out) coincided with the first period during which the plant operated—from December 10 to 16. Finally, further occurrences followed the reopening of the plant on January 10, 1969. (The one-month delay in the appearance of the next mortality was probably associated with movement of the causal toxin from the harbour out into the open waters of the bay.) The problem was not so much establishing the source of the trouble, as it was documenting a foolproof case that included a direct demonstration of the cause. To close down a multi-million dollar operation because of the possibility that it might be the cause of a fish kill is not something that a government, or, for that matter, any organization could do lightly, without fear of being subsequently sued, particularly when the actual cause of death remained unknown.

Research scientists established that effluent from the plant was toxic to animal life. Three potentially toxic substances were identified: acids, cyanides, and elemental phosphorus. Experimental tests showed only elemental phosphorus was capable of producing the "red herring."

The next step was to show that elemental phosphorus was actually present at the time of death of the fish in sufficient concentrations to cause mortality. This necessitated the development of a new method to analyze trace amounts of elemental phosphorus in fish tissues. After the method had been perfected, it was applied to normal and frozen specimens of fish collected at the time of the previous kills and to fish reared in the laboratory in the presence of known concentrations of elemental phosphorus. The results showed the livers of affected fish contained extremely high concentrations of elemental phosphorus, with lesser amounts in fatty tissues and still less in muscle. With this demonstration, it was obvious to the industry that a firm legal case was on hand and ready to be used. Disclaiming any responsibility, the management voluntarily decided to close down the plant on May 2, 1969, for the installation of pollution control facilities.

It is estimated that the total cost to the company due to closure of the plant, reimbursements to fishermen for their losses, removal costs for "vacuum-cleaning" the bottom of Long Harbour with a suction dredge, and other factors, was $5,000,000. The plant returned to operation again. The 450 kilograms per day of elemental phosphorus formerly discharged to the waters of Long Harbour each day were now recovered in plant operations or detoxified by oxidation prior to discharge. The phosphorus plant was closed permanently in 1992, and a final decommissioning order was issued in 1996.

Although industrial cases of human poisoning from elemental phosphorus were known from the early part of the 19th century, this was the first documented case of environmental effects. Toxicity of an element that is so essential to life is highly unusual. The Canadian investigating team unearthed only one previously published experimental study of the toxicity of elemental phosphorus to fish. It was subsequently learned, however, that a comparable case of environmental pollution with elemental phosphorus occurred in Sweden and had been documented in government reports (in Swedish) but was not widely publicized.

Government inspectors who originally approved the design of the factory at Placentia Bay were, unfortunately, ignorant of the harmful environmental effects that would ensue. The industry, well aware of the toxicity of yellow phosphorus to humans, assumed it would be rapidly oxidized in the natural environment and thereby rendered harmless.

However, it turned out that elemental phosphorus in colloidal form is not readily oxidized in water, a fact that could easily have been established by prior experiment. It was the death of fish that gave the warning signal to humans.

Were it not for the fish and the alertness of fishermen, fishery inspectors, and the research team that ultimately traced the cause, irreparable human damage could have been done. As little as 0.6 milligrams (0.00002 ounce in weight) of elemental phosphorus, an amount equivalent to a few grains of salt, can induce hepatitis in human. A dose of 100 milligrams (0.003 ounce) results in death.

There are also examples of phosphorus being toxic to humans in high doses or in very toxic forms. For example, phosphine, mentioned earlier as the gas causing *ignis fatuus* in swamps, is generated in grain bins where aluminum phosphide is used as a fumigant. The gas is extremely toxic. There is a documented case of a pregnant woman killed by accidental exposure to phosphine generated in stored grain.

With this, the chapter on "Phosphorus, The Morning Star" ends. Attention is focused next on phosphorus and the control of human-caused eutrophication.

# 6 | The Environmental Physician

*Methods of controlling cultural eutrophication by eliminating point sources of nutrients*

IN CHAPTER 2, it was suggested that the water transport system of waste disposal should be regarded as an important cause of eutrophication problems that appeared during the 20th century. Our typical way of disposing of sewage is really just a roundabout way of dumping plant fertilizers into water. The process starts with the application of fertilizers rich in phosphorus and nitrogen to soil, replacing the quantities removed in the harvest of crops. Generally, fertilizer is over-applied, because it is inexpensive when compared with the value of the crop. As a result, the excess leaks into groundwater, lakes, and streams. The harvested crops in turn serve as food for humans, leaving the body in the form of urine, feces, and the carbon dioxide of exhaled air.[1] Nutrients in urine and feces then flow via sewage collection systems into rivers, lakes, estuaries, and oceans, where they act as fertilizers that amplify the growth of aquatic plants. Introduction of phosphate-based detergents after World War II added another technological component to the one-way flow of nutrients to sediments and the sea. In the mid-20th century, human wastes and detergents accounted for over half of the total phosphorus reaching water in many areas. These were entering lakes largely from *point sources*, the pipes that released treated sewage. Since the mid-20th century, livestock culture and land fertilization have increased enormously as a response to increased global demand for food. Cities are growing rapidly, as rural populations migrate to urban areas, and the area of paved roads and parking lots increases. The runoff of nutrients from agricultural land, septic systems, urban streets, and lawns has increased to become important problems. These *non-point sources* of nutrients are more difficult to control. In this chapter, we shall describe the situation in the mid-20th century, when point sources were the predominant sources of nutrients. We shall treat the current situation, when non-point sources have often become more important than point sources, in Chapter 10.

Attention is first focused on early methods for control of cultural eutrophication. Some of these, such as the use of toxic chemicals to kill aquatic plants, merely remove the signs of eutrophication without attempting to treat the cause. Others "solve" the problem by displacing it to another area, as in sewage diversion schemes. In localities where property values are high and land is at a premium, the most reasonable approach may be to remove phosphate from sewage by chemical precipitation, using the phosphate-rich sludge as fertilizer on land. The most efficient solution in an ecosystem sense is the totally cyclic one, the use of sewage treatment plant effluents as fertilizers to increase the yield of some harvestable crop. In this case, however, care must be taken to prevent spread of disease and contamination of the cyclic food chain with industrial toxins. The increasing concentration of human populations in urban areas poses special problems. Food, with its nutrient content, is imported from watersheds far away to be concentrated in the few watersheds where people live. To be ecologically efficient and avoid overloading the watersheds and waters where people live, the resulting effluents should be transported back to areas where the food was grown. However, by diluting it with many gallons of water, contemporary sewage treatment makes this nearly impossible.

Costs associated with treatments to reduce eutrophication vary depending on the treatment, the population served (per capita costs declining with increasing population), and various economic factors. The cheapest method in a short-term sense is to poison unwanted plants with some aquatic herbicide; however, this is also the most dangerous method in terms of possible side effects and the most expensive in the long run, because it does nothing to remove the cause. Diversion schemes become undesirably expensive even for large metropolitan centres when distances increase beyond 25 kilometres (15 miles). Although construction costs may be high, depending on the local terrain, once a diversion has been completed, maintenance and operational costs are low. Operational costs for phosphorus removal, based on recycling or chemical precipitation of phosphate, typically range from $8 to $70 per person per year depending on the method, the removal efficiency and the size of the treatment plant. Local prices of chemicals, sludge removal costs, and other factors also contribute to the variation in cost. "A penny a day keeps cultural eutrophication away" was a reasonable slogan for most citizens of the United States and Canada in the early 1970s, although with recent inflation, "a dime a day..." may be closer.

Most of us would prefer not to have to pay for sewage treatment, thinking matters will always take care of themselves "some other way." When it comes right down to it, sewage treatment is not a terribly exciting business.

This tendency to view sewage treatment only in terms of costs is somewhat like going to a restaurant for dinner and then ordering solely on the basis of lowest price. When sewage treatment is examined in the broader context of health, environmental benefits accrued, and expenditures that would otherwise have to be made, it turns out to be surprisingly cheap. On a per-capita basis, the cost of nutrient removal is roughly equivalent to the purchase of 1 litre (1 quart) of bottled water a month.

There is a close functional similarity between the human body and human-dominated parts of the biosphere. Both are ecosystems in their own way: one controlled (in part) by the will of an individual, and the other is controlled (also, in part) by the will of a community. In both cases, it is human will that determines the behaviour and health of the system within limits imposed by nature. When the human body is injured or some part of the system goes out of control, the mind of the individual directs the body to a medical physician for help. In a like manner, when an ecosystem is injured or some part of it (a pest species, for example) goes out of control, the collective mind of the community calls an environmental physician for help—that is, an experienced ecologist. In each case, the physician does the best he can with the tools available, writing out a prescription, giving general advice, or perhaps suggesting the need for hospitalization. However, he cannot force the patient to do anything, even if the patient is dying. The patient has to will himself back to health.

This comparison of an ecosystem to the human body is appropriate. However, one qualification must be clearly understood. Expressed in terms of medical practice, our knowledge of ecosystems today is equivalent to that of the human body in the latter part of the 18th century. At that time, the internal anatomy of the body was only known in gross detail, and physiology was hardly understood at all, except for William Harvey's demonstration of the circulation of the blood. Bacteria had been seen but were not recognized as agents of disease; knowledge of vitamins, hormones, and antibiotics did not exist; and the need for aseptic procedures in surgical practice was not realized. This realization makes one regard the grandiose claims frequently made by politicians, developers, and cottagers' societies that they intend to "manage and enhance" our freshwaters seem absurdly arrogant.

In many cases where sewage or manure are the sources of the nutrients that cause eutrophication, lakes and rivers are also receiving pathogenic organisms. In the first edition of this book, Vallentyne (1974) wrote: "Based on this analogy, we should not be too surprised if, in the latter part of the 20th century, we are inflicted with environmental ills equivalent to epidemics of typhoid fever, cholera, and bubonic plague that characterized earlier times. In our ignorance, we will attribute such environmental catastrophes, when they happen, to forces beyond our control, as disease was regarded in the 18th century."

Very little has changed in the Third World since that time. Typhoid, cholera, and other waterborne diseases still kill hundreds of thousands annually. However, we also see problems with waterborne pathogens in the developed world that were not regarded as important in 1974. An outbreak of *Cryptosporidium*, a protozoan that infects humans, wildlife, and livestock, killed over 50 people and caused up to 400,000 cases of gastrointestinal illness in Milwaukee in April 1993. *Cryptosporidium* has also been a problem in other centres, including Edmonton, Alberta. This pest is resistant to chlorination, and to eliminate it, Edmonton has installed costly UV radiation treatment on its water intakes. The source of 70% of the pest is feedlots upstream of the city's water intake on the North Saskatchewan River.

In 1974, the common fecal coliform *Escherichia coli* was usually regarded as a warning signal that fecal matter was reaching water. The virulent strain O157:H7 was unknown but soon to appear. The O157:H7 strain is thought to have spread throughout the western world with imported cattle. Contamination of drinking water with O157:H7 killed seven people and sickened an estimated 2300 in tiny Walkerton, Ontario, in May 2000. For many other examples of drinking water contamination in North America, the reader is referred to Hrudey and Hrudey (2004). It is obvious that some of the sources of nutrients that cause eutrophication are also causing increased waterborne illness.

Another emerging problem connected with eutrophication is an increasing incidence of toxins released by blue-green algae blooms. There are numerous documented cases of liver failure in animals that drank from lakeshores where blue-green blooms were highly concentrated. The 2003 algal bloom on Lake Winnipeg that covered 6000 square kilometres released toxins that were many times the World Health Organization's recommended limits for human consumption, these words seem strangely prophetic, Truly, we are in the midst of the Algal Bowl!

To environmental physicians who will look back on us from the vantage point of the 23rd century, it will be obvious that, in the 20th century, ignorance of the causes and an underestimation of the severity of the results marked much of what has happened to freshwater. They will say we lived at a time when forces of the environment were not understood. This was certainly true through the 1960s. Since that time, there have been many attempts to correct earlier problems, without which our waters would be much more severely degraded than they are. Much of the science that we need to prevent and correct many problems is also known, and we have made some progress, with legislation and financial resources to counteract the degradation of lakes. Some will claim that we have made outstanding progress, However, with increasing populations, livestock, fertilizer use, and industry and with a warming climate, we are still losing the global battle against eutrophication.

The environmental physician may advise his patient to stop smoking (abate air pollution), go on a diet (control population), slow down in activity (reduce the pace of technology), or drink in moderation (avoid excessive consumption of hydrocarbons), but he cannot force the patient to conform. Whether the patient conforms is up to the patient, not the physician. If the patient follows the physician's advice, it is probable, although by no means automatic, that improved health will result. On the other hand, if the mind controlling the body behaves like an alcoholic, there is little the environmental physician can do other than attempt to assist through education.

On this note of realism, some prescriptions used in the treatment of cultural eutrophication may now be examined. They are given in approximate order of first historical use (except for recycling, which is the oldest of all). The first prescriptions deal with some "quack" medicines for "Treatment of Symtoms in Lakes" (headache tablets for appendicitis might be an appropriate medical analogy). Attention then shifts to "Diversion of Sewage Wastes" (how to pass your schizophrenia downstream). Then, following a general account of the significance of nutrients in the control of eutrophication, responsible remedies (ones that deal with causes) are discussed under the following headings: "Removal of Nutrients from Sewage;" "Nature (Recycling) for Nutrient Removal;" and "Control of Agricultural Wastes." The sequence ends with a history of Lake Washington, USA, which shows the critical role of phosphorus in a sewage effluent diversion scheme.

### Treatment of Symptoms in Lakes

ONE OF THE EARLIEST METHODS to control algae and aquatic weeds in ponds and lakes was adding chemicals that inhibit growth. This approach made no attempt to come to grips with the causes involved. It is like the man who solves the problem of a leaky roof by placing a pan underneath to catch the drip.

One of the more common treatments to reduce the incidence of blue-green algae blooms used to be the application of copper sulfate to surface waters. Copper in ionic ($Cu^{2+}$) form is a general biological poison; however, by judicious application in relatively low concentrations, it can bring about the death of algae without detriment to fish or aquatic invertebrates. This was the method used between 1912 and 1958 to reduce the incidence of blue-green blooms in the chain of lakes at Madison, Wisconsin, discussed in Chapter 2. Thousands of tons of copper sulfate were used over a 46-year period in what eventually was recognized as a futile effort to create clear-water conditions in lakes still polluted with nutrients.

Most of the copper added to the Madison lakes precipitated almost immediately on contact with water in the form of insoluble compounds produced by reaction with carbonate and bicarbonate ions naturally present in the water. The copper now lies locked in the bottom sediments of the affected lakes. By no stretch of the imagination could the use of copper sulfate to reduce algal growth be considered as anything other than a temporary control. Effect of a single application in terms of preventing algal blooms seldom lasted for more than a few weeks.

Another chemical used for the temporary treatment of both natural and human-caused eutrophic bodies of water is alum (potassium aluminum sulfate). Unlike copper, alum does not act as a toxicant but, rather, as a chemical coagulant. It forms a flocculent precipitate of aluminum hydroxide that acts as a trap for phosphate ions and organic materials. These are removed from lake water with the settling of the precipitate. The presence of aluminum hydroxide at the sediment surface may also help keep phosphate in the sediments of shallow ponds and lakes. Since alum acts as a coagulant for materials in water (rather than as a poison) it is of no use in streams where a fresh supply of nutrient-rich water enters every second. It can also cause extreme acidification when used in the poorly buffered soft waters of eastern Canada and adjacent parts of the United States. More recently, application of iron has proved to be somewhat effective, by accelerating the precipitation of phosphorus, as discussed later in the book.

Other chemicals used in the mid-20th century to control the growth of algae and aquatic weeds include compounds of arsenic and mercury, 2,4-D, and a wide variety of organic compounds with suspiciously short abbreviations for long technical names. Some of these are now known to be dangerous nonspecific and persistent toxins that can have severe effects on fetuses and newborn infants. Many others have not been adequately tested for effects on ecosystems. The use of chemicals to control algae is now much less common than it was 30 years ago, so humans seem to be learning slowly.

One pollutant-free method purported to reduce blue-green blooms in lakes is controlled circulation and aeration of water by introducing a stream of air bubbles below the thermocline. The destratification of lakes prevents anoxia from forming in deep water, and by circulating the entire water column, it keeps bloom-forming species from concentrating at the lake surface. As a result of phytoplankton spending much of their time deeper in the water column, light is more likely to limit photosynthesis. The principal virtue of the method is that it does not involve the addition of foreign chemicals, but its effectiveness is highly variable. The pumps necessary to aerate or circulate water are also very expensive. Most use precious fossil fuels, and release greenhouse gases, although recently solar and wind-powered units have begun to appear. The technique has only proved feasible in small to medium-size lakes, and even there the effectiveness is variable. Recently, it has been proposed that wind generators or solar panels might be used to power these devices, but it remains to be seen whether they will be effective.

Mechanical weed cutters, the aquatic equivalents of combine harvesters, have been used for control of rooted aquatic plants. For maximal effect, the harvest should be deposited on land under circumstances that favour absorption and retention of nutrients by soil, preferably outside the basin of the lake to be protected. Merely burning the dried remains at the lakeshore is insufficient to solve the problem, since phosphorus and nitrogen salts are concentrated in the ash, simply returning to the lake with runoff. Again, mechanical weed harvesting is expensive and releases fossil fuels. Often, the weedbeds removed are critical habitats for fishes.

### Diversion of Sewage Wastes

SEWAGE DIVERSION collection systems have been constructed in some areas to prevent nutrient-rich effluents of sewage treatment plants from entering bodies of waters used for recreation or water supply. Lac d'Annecy in France, the Lake of Zürich in Switzerland, Schliersee and Tegnersee in Germany, Lake Washington, and the Madison lakes in the United States are lakes that have been protected in this way.

Diversion, of course, is not treatment. It merely passes the problem downstream. Provided no one is living there, and nutrients do not cause a problem at another site, it can be advantageous. Often, however, it is only a momentary solution in terms of population expansion and land development.

Seas and oceans are prime areas for waste disposal from an engineering point of view because of their immense volume and absence of human inhabitants. Also, in many areas of the oceans, an increase in productivity would be beneficial. However, even marine waters can be adversely affected by nutrient enrichment, particularly areas with restricted circulation such as estuaries and embayments. The Baltic Sea, Chesapeake Bay, the Gulf of Mexico, the Adriatic Sea, and many other coastal marine areas are showing signs of eutrophication as a result of the addition of nutrients and organic wastes from human populations surrounding their shores.

Entire continental shelf areas of the oceans are affected. We shall discuss this marine eutrophication in more detail in Chapter 13.

The history of sewage treatment and the location of effluent pipes in Madison, Wisconsin, outlined in Chapter 2 is the best commentary on diversion schemes. The first diversion merely shifted problems from the uppermost lake in a watershed to the lake below. The second diversion transferred the problems to the two lowermost lakes in the chain. The third diversion transported sewage effluent around all four lakes via Badfish Creek to the Yahara River. The "trade-off" has resulted in an overall benefit. However, it took the citizens a long time to understand the difference between diversion and cure. There may still be more to learn.

One interesting case combines diversion with reuse of both the water and nutrients in sewage. It shows what can be done when a little human ingenuity is applied to a suitable geographic terrain.

Mexico City is built on the site of the ancient Aztec city of Tenochtitlan, which is estimated to have had one million citizens when conquered by the Spanish in 1519. It lies in a high basin, elevation 2200 metres, surrounded by the high peaks of extinct volcanoes. It is one of the world's largest metropolitan areas, estimated to have a population of over 20 million people, growing at about 350,000 per year. Originally, the city was essentially an island in the middle of a lake, which provided protection from enemies, as well as water for drinking and agriculture. Aztecs used muds from the lakes in the basin to construct terraced *chinampas*, where water from growing crops was supplied by an intricate system of aqueducts and canals. Howeer, the Spanish invaders began to drain the lakes for pastures and eliminate aqueducts for roads. By 1850, it was necessary to begin drilling deep wells to tap groundwater in the basin. Most of the city's water is now taken from over 1000 groundwater wells, at a rate of 55 cubic metres per second. The original lakes have all but disappeared, and the aquifers are replenished at only 28 cubic metres per second. As a result, groundwater levels are declining rapidly, and the clay soils of the area are shrinking and cracking. The centre of the city has sunk by as much as 9 metres and continues to sink by several centimetres per year. The subsidence has caused many problems with leakage in sewage and water supplies, as well as in the foundations of buildings and historic monuments. Although all of the city's drinking water is treated, 585 of every 100,000 residents die annually from waterborne diseases, evidence that the distribution system is contaminated.

Sewage from the area was once removed by gravity, but because of the lowered ground level, this is no longer possible. Some of the sewage still travels in an ancient grand canal, but sinking ground levels now makes it necessary to pump water into the canal. There is also an extensive network of deep drainage pipes to transport sewage. In the 1990s, 68 pumping stations pumped sewage through 10,000 kilometres of pipe and 111 kilometres of open channels to remove it from the city. While 82% of the population is connected to the sewage system, over 90% of the sewage is untreated. A cadre of full time "sewage divers" grope through the black waters of the pipes to maintain the shaky infrastructure, occasionally finding human bodies or even automobiles clogging the system...certainly one of the world's most unpleasant and dangerous occupations! Untreated wastes from the city are eventually pumped into the Grand Drainage Canal that drains via the Tequisquiac Tunnel to the Montezuma River, then to arid and sparsely populated receiving basins, where the inhabitants want both the water and nutrients (the "black water," as they say) to make the most of unfavourable agricultural land, most degraded by overgrazing with sheep. Much of the untreated water is used to irrigate cropland. Over 80,000 hectares of land in the State of Hidalgo are irrigated with the untreated sewage. At first glance, this seems like an excellent plan for recycling nutrients. However, contamination from pathogens, from trace metals, and salinization of soils have confounded the plan. Bouts of cholera and other gastrointestinal diseases are common.

The other 28% of Mexico City's population discard wastes into ditches, nearby surface waters, or onto the streets. At present, Mexico City's water and sewage systems are considered to be in a critical state, unable to meet the water demands of its growing population or adequately treat their wastes.

### Significance of Nutrients in the Control of Eutrophication

METHODS FOR CONTROL of eutrophication discussed up to this point make no attempt to reduce the delivery of nutrients derived from human culture to water. As attention now shifts to methods involving nutrient removal and recycling through soil, a preliminary account of the relative importance of phosphorus to other nutrients in the control of eutrophication may be helpful.

In Chapter 1, phosphorus and nitrogen were identified as the two most important nutrients involved in the triggering of plant growth in lakes.

Other things equal, it would seem likely that one or the other would best serve as a means of control.

In 1887, S.A. Forbes's classic paper, "The Lake as a Microcosm," described lakes as ecosystems separated by sharp boundaries from their surrounding terrestrial ecosystems. Lakes were regarded as microcosms for much of the 20th century. The importance of nutrient sources from "outside" a lake to eutrophication was not fully realized until Richard Vollenweider's classic review of the problem in 1968. The impact of his work will be treated in detail in Chapter 9.

The primary reason for attaching more importance to phosphorus than nitrogen (or for that matter, any other plant nutrient) is that of all chemical elements required by plants, phosphorus is the most controllable (by man) and the most growth-controlling in terms of plants. The relationship is not accidental. It results from the use of the same controls by nature and man. Both are due to the geochemical abundance of calcium, iron, and aluminum ions and to their tendency to form water-insoluble precipitates with phosphate ions. The rationale for phosphate removal to control cultural eutrophication is as follows:

1.  Laboratory and field experiments have shown that chemical treatment to remove phosphate eliminates most of the fertilizing effect of effluents from sewage treatment plants on natural waters. The demonstration that this effect can be reversed on the re-addition of phosphate alone shows that it is due to removal of phosphate, rather than some other factor.

2.  Technology for removing phosphate from sewage is readily available and economically feasible. Technology for removing other nutrients, including nitrogen, has not been developed to the same extent and is more costly.

3.  In most lakes not polluted by man, phosphorus tends to limit plant growth more than any other element. Thus, by restricting the supply of phosphorus to lakes, natural controls are utilized to the fullest extent.

4.  In culturally eutrophied lakes, the proportion of the total supply of phosphorus directly attributable to humans is typically higher than for any other growth-limiting element. In other words, phosphorus is more controllable because more of it originates from humans and their activities.

5. Phosphorus is retained by soil to a greater degree than nitrogen. By accentuating this natural control process with the removal of phosphorus from sewage, a maximum return on the investment is almost automatically assured.

6. When the supplies of readily available phosphate in water are depleted to the point that they limit further growth of plants, no accessory supplies are available from the atmosphere. Nitrogen, on the other hand, can be transported across the air–water boundary both as elemental nitrogen ($N_2$), a major constituent of air, and as ammonia ($NH_3$). Most organisms are unable to use elemental nitrogen; however, certain blue-green algae and bacteria can. They convert elemental nitrogen into ammonia and organic nitrogen compounds, which then become available as nutrients in aquatic ecosystems. Unfortunately, many nitrogen-fixing blue-green species also produce unsightly surface scums that are not readily used by herbivores. In some cases, these produce chemicals that affect the taste and odour of water, and are toxic to vertebrates, including man. We shall describe these important details of the eutrophication process in Chapters 9 and 10.

These six points form the central basis of the rationale for the removal of phosphate from municipal and industrial wastes as a means of controlling cultural eutrophication. As we shall discuss later, supplies of phosphorus from agricultural sources must also be reduced in areas where there is an appreciable transfer from land to water. Where these wastes originate from confined feeding operations (CFOs), the same techniques used for human sewage would work. It is, in fact, always wise to determine the relative quantities of phosphorus arising from various natural and human sources. If the major source is from agriculture, treatment of human sewage for phosphate removal may have little effect in curing eutrophication.

The first question often arising in the minds of those unfamiliar with aquatic ecosystems is whether a reduction in the supply of a single nutrient, such as phosphorus, can control accelerated plant growth in the aquatic environment. Some people believe eutrophication is so complicated and poorly understood that it is naïve to think control via any single element, such as phosphorus, could work in the majority of cases. Sunlight, nitrogen, carbon, and a variety of other factors are known to limit growth of aquatic plants under natural conditions. Proponents of such arguments, however, show by the very questions they raise that they do not understand the distinction between causes and mechanisms of control.

Two examples will reduce the distinction to its utmost simplicity. The first is that it is always easier to destroy than it is to build—the means of destruction having little to do with the complexity and means of construction. The second concerns Asiatic cholera, a disease caused by the bacteria *Vibrio cholerae*, which was rampant throughout the world prior to the 20th century. The discovery of an effective control mechanism for cholera was made 30 years prior to the discovery of the causal organism by Robert Koch in 1883. The mechanism of control was deduced without knowing anything about the cause of the disease or the manner in which the cholera bacillus affects the human body. The secret was to separate sewage from the drinking water supply. Modern water and sewage treatment have all but eliminated the cholera problem in North America and Europe, but it continues to be an enormous problem in Asia, Africa, and Latin America.

One of the principal reasons limnologists are confident that cultural eutrophication can be reversed is the fact that individual applications of phosphate-rich fertilizers to lakes to increase fish production generally produce only short-lived effects. The situation is similar to the effects of alcohol on the human body at a cocktail party; the system is disturbed for a time, but soon returns to its previous state. Carrying the analogy further, cultural eutrophication is like alcoholism, as it is the continued supply of the disturbing influence that creates and perpetuates the disturbed state. If the will to cure exists, it can be cured.

In summary of the information presented above, it can be said authoritatively that, in most freshwaters, attempts to control eutrophication without restricting the supply of phosphorus are likely to be ineffective.

## Removal of Nutrients from Sewage

THREE PRINCIPAL METHODS are available for removing phosphate from sewage by chemical precipitation. These are based on reaction with lime (calcium oxide), iron (in various forms), or aluminum (usually as potassium aluminum sulfate or alum). All of these methods have the added advantage of removing organic substances and often some nitrogen as well as phosphate.

Relative costs of the different treatments vary with local prices of chemicals, the particular engineering treatment used, the need for capital outlays, and so on. Sometimes, a waste can become a resource, as in the case of waste "pickle liquor" from the steel industry. This liquid waste, rich in acid and iron, can be combined with sewage to remove phosphate, acid, and iron. In the process, a waste is transformed into a resource.

One factor that can be of concern is sludge disposal. This is particularly true in the case of the lime method because of the large amount of sludge caused by precipitation of calcium carbonate. Precipitation methods depend on the addition of chemicals to sewage. The added calcium, iron, or aluminum ions end up in sewage sludge, and the sulfate ions that typically form the other part of iron and aluminum salts pass into water. Trace metals and organic molecules, some of them toxic at high concentrations, are also removed with the sludge. Although this procedure appears to be adding to pollution, it is really trading off a serious problem for a lesser problem.

Drs. G.V. Levin and Joseph Shapiro developed a method for the removal of phosphate from sewage that does not depend on the addition of chemicals. The method is based on the knowledge that microorganisms associated with sewage sludge take up excess phosphate under well-oxygenated conditions and release it in the absence of oxygen. There have been many more advances. For example, many modern sewage treatment plants expose effluents to high-intensity UV radiation before releasing it into lakes or rivers. This treatment kills almost all of the pathogens.

In terms of control of eutrophication, nothing further is gained by treating sewage for removal of nitrogen compounds if phosphate has already been removed. It takes only one growth-controlling element to curtail plant growth. Nitrogen, however, may constitute a problem in its own right in contributing to dangerously high nitrate levels in water or in consuming dissolved oxygen, as in the oxidation of ammonia to nitrate.[2] At the start of the 21st century, nitrogen removal has become necessary in parts of the midwestern United States, where nitrate leached from fertilized cornfields and massive hog farms has contaminated ground and surface waters. In soft-water areas, ammonia released from agriculture and industry can be oxidized to nitrate, a process that acidifies soils and poorly buffered freshwaters.

Various processes are available for removing nitrogen, based on distillation of ammonia at high pH, biological denitrification (conversion of the nitrate formed after ammonium is oxidized to elemental nitrogen that then passes to air), or removal by ion exchange resins similar to the zeolites used in water softening. Methods that involve "stripping" of ammonia from water into air only redistribute the problem, unless the ammonia is removed and converted to elemental nitrogen before it is released.

## Nature (Recycling) for Nutrient Removal

NUTRIENT REMOVAL by addition of sewage or treated sewage effluents to soil, the most ancient method of waste disposal, is still used in many parts of the world (though rather rarely compared to more modern forms of sewage treatment). Nevertheless, it is the best illustration of a total ecosystem approach and, as such, is being rediscovered in some parts of the United States and Canada today. The community of South Tahoe (bordering Lake Tahoe in the southwestern United States) pumps the effluent from its $19,000,000 sewage treatment plant over a drainage divide into a reservoir, where it is fed out for irrigation. In some parts of Pennsylvania and Michigan, nutrient-rich effluents from sewage treatment plants are used to promote forest growth. In the dry Okanagan Valley of British Columbia, sewage is being used on an experimental basis to accelerate production of grain. In principle, these practices do not differ from the use of animal manures for fertilization of agricultural land. However, human waste has many features that natural animal wastes do not. It contains dozens of foreign chemicals. Toxic trace metals, such as mercury, lead, and zinc; hormones from birth control pills; naproxen and salicylic acid from pain killers; antibiotics; caffeine; and chemicals from cosmetics are a few of many examples. Some of these chemicals pass through the human body and are excreted, but it is also still common for people to dispose of unused medicines or household chemicals by dumping them in the toilet. Many of these substances are not completely removed during sewage treatment. Although concentrations of individual chemicals are low, their combined effects are not known. In one disturbing study, Dr. Karen Kidd and her colleagues (2007) found that small concentrations of the estrogen found in birth control pills, 17α-ethynylestradiol, caused feminization of male fathead minnows (*Pimephales promelas*) in an experimental lake. The fish population nearly collapsed as a result of decreased reproduction. It is suspected that small amounts of antibiotics may lead to the development of antibiotic resistant bacteria, which are now common in most hospitals. The increasing use of chemicals in animal culture is now causing contamination via animal wastes as well.

There is also some concern about pathogenic organisms that are not destroyed in sewage treatment. When applied to land, they can be washed into drinking water supplies. In brief, although recycling human excrement seems like an excellent idea in theory, the process requires careful monitoring and control in practice. There are some recent advances. For example, high temperature composting and careful separation of sewage from industrial wastes are being used in some areas to prevent problems, but such measures are still not in widespread use.

Wastes of society can be put to good use in accelerating the growth of agricultural or forest crops if human pathogens or toxic pollutants are not redistributed in the process. This approach finds its best application in tropical or semitropical areas, where plant growth occurs throughout the year. In latitudes lying closer to the poles, its application is more limited because frozen soils inhibit decomposition in winter, and wastes spread on frozen land are often transported to waterbodies before soils thaw.

Under some circumstances, standing bodies of water can also be used for nutrient removal when sites are available for the purpose and climate permits year-round growth of aquatic plants. In warm climates, sewage lagoons and carp (*Cyprinus carpio*) ponds often become one and the same thing, with carp transported to clean water several weeks before harvest. In some parts of Southeast Asia, it is not uncommon to find outhouses and poultry pens located directly over carp ponds. Wastes from one "industry" become directly available as a resource for the next. In a series of specially constructed ponds, the city of Munich, Germany, puts its wastes to use in the production of fish.

In temperate latitudes, sewage lagoons do not function as well in terms of nutrient removal. They are too shallow and, typically, are flushed out so rapidly that they behave more like rivers than lakes. Short flushing times and inadequate depths do not permit burial and retention of nutrients in sediments. They are also ineffective in terms of nutrient removal during winter when the reduction of sunshine limits algal growth. They can even serve as a source of further nutrient supply by releasing nutrients previously deposited in the sediments.

There are also cases where wetlands have been successfully used for reducing nutrient concentrations in effluents. The higher density of plants than in lagoons or ponds can cause higher rates of nutrient removal. However, they, too can become nutrient saturated, they are not effective in winter, and their use must be carefully monitored.

In summary, there are ways to utilize either natural or constructed ponds or wetlands to reduce the amounts of nutrients, silt, pathogens, and toxins that would otherwise enter lakes, rivers, and groundwater, some of which we want to utilize as sources for drinking water and other purposes. The value of such systems is only beginning to be recognized.

## Control of Agricultural Wastes

IN MOST COUNTRIES, physiological wastes produced by farm animals outweigh those produced by humans by a factor of 10:1. As we shall see in Chapter 10, in some western states and provinces that are heavy livestock producers, the ratio can be 30:1 or more. In the mid-20th century, it was believed that most animal wastes were delivered directly to soil, where they would be recycled into crops and that eutrophication in agricultural areas would not be a widespread problem However, the explosive growth of livestock culture since that time has brought eutrophication even to areas where human populations are low.

Another problem inherited from agriculture arises from economically wasteful fertilization practices. In some areas, manures and artificial fertilizers are applied to soils in late autumn or even in winter on top of frozen ground. Depending on the rate of melting of snow in spring and the slope of the terrain, major fractions of nutrients can be carried away in solution or attached to eroding soil particles. The circumstances differ little from the runoff of water over cement.

Inputs of nutrients to water from agricultural and livestock-rearing operations can be enormous. Intense fertilizer use and large feedlots contribute much of the nutrients delivered to water in the midwestern United States and parts of southern Canada. In addition to nutrients, pathogens, antibiotics, and hormones originate from agricultural use, posing problems for treatment of drinking water and for receiving ecosystems as discussed above. The treatment of agricultural wastes is still in its infancy, and there is a need for great improvements to be made.

## History of Lake Washington

ONE CASE of sewage diversion has been of considerable importance in interpreting the role of phosphorus in the control of eutrophication. This is the well-documented limnological history of Lake Washington, a body of water 128 square kilometres (50 square miles) in surface area with a maximum depth of 59 metres (194 feet) and a mean depth of 18 metres

FIGURE 6.1: *Satellite image of Lake Washington showing the intense development in its catchment by the city of Seattle. For colour image see p. 329.*

Image from NASA Landsat Program, 2002, Landsat TM scene LT5046027000220210, Orthorectified and Terrain Corrected. USGS, Sioux Falls, SD, 07/21/2002.

(59 feet). It lies within the boundaries of metropolitan Seattle, Washington, USA.

Most of the problems of Lake Washington have been associated with population increase and associated changes in the flow of sewage effluents. The population of metropolitan Seattle was about 300 in 1865 and rapidly increased to over 1,200,000 by 1965 (Figure 6.1). The first growth of the city occurred between Lake Washington and Puget Sound, the latter an arm of the Pacific Ocean. In 1922, pollution problems created by 30 outfalls of raw sewage to the lake were so bad that a diversion was created to carry sewage to the sound. The lake soon reverted to a healthy state.

In 1930, sewage effluents began entering the lake from treatment plants in outlying communities. By 1962, the discharge of these effluents had risen to a level of 76,000,000 litres (20,000,000 U.S. gallons) per day. In all cases, the sewage was given normal treatment for removal of organic substances but not for removal of inorganic plant nutrients.

By the mid-1950s, it was obvious to Dr. W.T. Edmondson, noted limnologist at the University of Washington, that the lake was beginning to repeat the history of eutrophication in the Lake of Zürich, Switzerland. *Oscillatoria* (now *Planktothrix*) *rubescens* and other blue-green algae started to bloom. Transparency of the water in midsummer, as measured by the visibility of a Secchi disc, decreased from a value of 3–4 metres (10–14 feet) in 1950 to less than 2 metres (6 feet) in 1957. In the 1950s, Dr. Edmondson began alerting city officials of other changes that would take place if nutrients continued to be added to the lake. The officials were concerned.

In 1958, citizens of metropolitan Seattle decided by a special vote to do something to save the lake. A diversion scheme to carry the nutrient-rich effluents to the sea was initiated in 1961. By the end of 1963, effluents discharged to the lake had decreased from 80,000,000 to 58,000,000 litres (20,000,000 to 14,500,000 gallons) per day. The values were reduced further to 44,000,000 litres (11,000,000 gallons) per day in 1965, to 800,000 litres (200,000 gallons) per day in 1967, and to zero in 1968.

As these changes were taking place, Dr. Edmondson and his students were on the lake measuring concentrations of various chemicals and kinds and abundance of algae and other microorganisms. Even before the diversion was half completed, evidence that the lake was improving began to accumulate. Year-to-year trends in the concentration of dissolved phosphate during late winter were beginning to show a decline, with a parallel decrease in the abundance of planktonic algae in midsummer. Transparency of the lake in summer was beginning to increase.

By 1970, Lake Washington was back to where it had been in 1950 in terms of plant growth and other signs of eutrophication. A rapid and remarkable recovery had taken place. Citizens and officials of the city were delighted with the result. By acting in time, they averted what would otherwise have been an embarrassing and undesirable situation, if not community disaster.[3]

The most significant finding, obtained by Edmondson and his associates at the University of Washington, was a direct correlation between concentration of dissolved phosphate in late winter and the abundance of planktonic algae in midsummer. None of the other nutrients measured showed a change comparable with the change in algal abundance.

The implications were clear. The primary effect brought about by diversion of wastes was the decreased supply of phosphorus to the lake. This, in turn, caused the reduction in plant growth. No other explanation fitted so well with the facts. However, after 1976, water quality in Lake Washington surpassed that in the earliest studies. This is now known to be the result of changes to the food web of the lake. The important role of food chain relations in eutrophication was scarcely known at the time. We will discuss it in more detail in Chapters 9 and 10. Other cases of recovery from eutrophication will be discussed in later chapters.

Nothing has been said in this chapter of the significance of detergent phosphates in eutrophication, yet in 1970, 50% or more of the phosphorus in municipal sewage in Canada and the United States arose from this source.[4] The next chapter discusses the controversy in 1969–1970 over the significance of detergent phosphates in the cause and control of man-made eutrophication.

# 7 | Detergents and Lakes

*The history of the 1969–1970 detergent phosphate controversy,
with an account of the role of phosphates in cleansing and how
their use was reduced*

## Pollution of the Lower St. Lawrence Great Lakes

ON OCTOBER 7, 1964, the Governments of the United States and
Canada requested that, under the Boundary Waters Treaty of 1909, the
International Joint Commission (IJC) inquire into and report as soon as
practicable upon three questions pertaining to pollution in the lower St.
Lawrence Great Lakes (lakes Ontario and Erie). The IJC, a permanent body
of three representatives from Canada and three from the United States, was
created in 1911 as an outgrowth of the Boundary Waters Treaty. Since that
time, its members have resolved a number of important problems along the
8763 kilometre (5443 mile) border to the mutual benefit of inhabitants on
both sides.

Avoiding any word or phrase that could possibly be construed as bias,
the two governments asked the IJC:

1.  Are the waters of Lake Erie, Lake Ontario, and the international
    section of the St. Lawrence River being polluted on either side of

the boundary to an extent which is causing or is likely to cause injury to health and property on either side of the boundary?

2. If the foregoing question is answered in the affirmative, to what extent, by what causes, and in what localities is such pollution taking place?

3. If the Commission should find that pollution of the character just referred to is taking place, what remedial measures would, in its judgement, be most practicable from the economic, sanitary and other points of view, and what would be the probable costs thereof?

Early in 1965, the IJC established two Advisory Boards to supervise the necessary investigations, one for Lake Erie and the other for Lake Ontario and the international section of the St. Lawrence River. The Boards were composed of officials from the state, provincial, and federal governments involved, who in turn drew upon their various resources to carry out the long and tedious investigations. The Boards' assessment was completed with the publication of a report containing a summary of findings and recommendations in November 1969, followed by two thicker and more detailed volumes of data in the spring of 1970.[1]

Among the 19 specific recommendations pertaining to various aspects of pollution in the lower Great Lakes, one created more discussion than all the others combined. That recommendation was for a program of phosphorus control to halt the growing trends of cultural eutrophication.

The Boards recommended that (a) phosphates in detergents be immediately reduced to minimum practical levels, with total replacement by environmentally less harmful compounds by 1972; (b) the remaining phosphate in municipal sewage effluents be reduced by not less than 80% prior to specific dates from 1972 to 1978 for different parts of the basins, with continued reduction thereafter to the maximum extent economically feasible; (c) programs be developed for the reduction of phosphorus inputs to water from agricultural sources; and (d) any new and significant changes in the addition of phosphorus to waters in the basins be regulated. The first recommendation created all the controversy, primarily because of the initially hostile reaction from the detergent industry. The recommendation was made because detergents, in 1969, accounted for more than half the phosphate in municipal sewage, a state of affairs that had come about since 1947. Manufacturers had been aware that governmental pressure was mounting for the removal of phosphates from detergents, but the timing and source of this recommendation came as a complete surprise.

There were good reasons for approaching the detergent question separately from the sewage treatment issue, as the IJC subsequently recognized in its final report. First, the removal of phosphates from detergents would result in a relatively rapid and permanent removal of at least 50% of phosphorus in municipal wastes. Secondly, it would result in a substantial reduction of treatment costs for phosphate removal at sewage treatment plants. (In the technical report to the IJC, annual savings in chemical costs associated with treatment in the lower Great Lakes basin were estimated to be $22,000,000 per year.) Thirdly, removal of phosphates from detergents would eliminate 50% of the phosphate delivered to water from isolated dwellings and small communities, situations in which any other kind of control would be virtually impossible to attain. Needless to say, benefits of such action would not be restricted to the lower Great Lakes basin.

Simultaneously with the release of the Summary Report from its Advisory Boards, the IJC arranged a series of public hearings in six communities spread around the lower Great Lakes drainage basin. The purpose of the hearings was to determine the reactions of other segments of society to the Advisory Boards' report. Hearings were scheduled for late January and early February 1970, allowing adequate time for any affected parties to make their own assessment of the report.

A different set of hearings on a closely related subject took place on December 15 and 16, 1969, in Washington, D.C. The Washington hearings were called by a Subcommittee of the U.S. House of Representatives Committee on Government Operations, headed by the Hon. Henry S. Reuss (Democrat, Wis.). The purpose of the hearings was to determine why the federal agency involved in the U.S. Joint Industry/Government Task Force on Eutrophication had been ineffectual in accelerating a search for suitable replacements for phosphates in detergents. So far as Mr. Reuss could see, the only product of 3 years of "intense" governmental effort of the Task Force had been a research grant for $99,896, the development of a provisional algal assay procedure, and some talk of a movie on eutrophication.

One of the more revealing parts of the testimony, presented at the hearings of the Reuss Committee, was a series of questions Mr. Reuss directed to Mr. Carl L. Klein, Assistant Secretary of the Interior for Water Quality and Research during the early part of the Nixon administration:[2]

*Mr Reuss:* Isn't it a fact that only 15% of the people of this country are now served by advanced waste treatment plants?

*Mr Klein:* I don't think that it is even that high yet.

*Mr Reuss:* Is it less than that, then?

*Mr Klein:* Yes.

*Mr Reuss:* And by and large the phosphate which shows up at sewage disposal plants comes from two main sources, does it not—household detergents and human wastes?

*Mr Klein:* Yes, sir.

*Mr Reuss:* And household detergents are made by three major manufacturers?

*Mr Klein:* That is correct.

*Mr Reuss:* And human wastes are made by a couple [of] hundred million manufacturers; is that correct?

*Mr Klein:* Yes, sir.

*Mr Reuss:* Well, doesn't it occur to you that it is easier to do something about three than about a couple [of] hundred million?

Mr Klein agreed but went on to say that, regardless of the detergent question, facilities would still have to be installed in major cities for removal of phosphates arising from physiological wastes. Mr Reuss concurred, but pointed out the time lag, because of the people involved, and the need for "getting phosphates removed from where most of them come from."

In the brief that the U.S. Soap and Detergent Association presented to the Reuss Committee (later repeated at the hearings of the IJC, the recurrent themes were the essential role of phosphates in cleansing action and the complexity of the process of eutrophication. The question was not whether people could have clean clothes *and* clean lakes, but whether North American society could maintain its standards of cleanliness and sanitation without any phosphate in detergents.

The culmination of the detergent issue came during the first week of February 1970, at hearings of the IJC, when high-ranking officials from federal water pollution control agencies on both sides of the border announced their intention to eliminate phosphates from detergents. When the big three detergent manufacturers and their phosphate suppliers heard these words, they knew their case was lost. For the second time in a decade, the detergent industry had been identified as an international polluter: first, from foams due to non-biodegradable detergents that transformed some areas of nature into what seemed like environmental washing machines, and now from phosphates as related to cultural eutrophication.[3]

The impact of the pronouncements was mildly dramatic, as the public slowly began to realize a major conflict existed between the soap and detergent industry and the two governments. In the resulting confusion, representatives of the press, radio, and TV scurried to ecologists to find out what eutrophication meant. Housewives, who had taken care to buy detergents with a *biodegradable* label on the box learned to their surprise in 1970, that the phosphate issue had nothing to do with biodegradability; non-biodegradable surfactants had been replaced years before. Few citizens realized that heavy-duty laundry detergents (those used in automatic machines) and automatic dishwashing detergents contained phosphates, let alone why phosphates were there or their effect on lakes and streams. Some public anti-pollution groups, however, were well aware of what phosphates and eutrophication meant and started campaigning for laws to take phosphates out of detergents.

## A Front Page Controversy

WHILE ALL THIS WAS GOING ON, other events were taking place behind the scene. Manufacturers and their phosphate suppliers were pressuring the two governments to avoid precipitous action that might throw the industry into chaos. Manufacturers stated that phosphates were essential for the cleaning power of household laundry and automatic dishwashing detergents. They claimed that, in spite of an intense search, not a single material or combination of materials had been found that could fulfill all the vital functions of phosphates, with no adverse effect on water quality. Unknown to the two governments, support for one aspect of the industrial stand was materializing from an unexpected source.

Two Canadian journalists, Robert F. Legge and Douglas Dingeldein, lashed out at the IJC, its Advisory Boards and a member of the Editorial Committee in a series of articles in *Canadian Research & Development* entitled "We Hung Phosphates Without a Fair Trial." (Figure 7.1).[4] With a crash self-education course on lakes and eutrophication, the authors claimed to reveal for the first time "a new account of a principle process of eutrophication that shows carbon, not phosphorus, to be the controlling nutrient in the production of algal bloom."

The scientific basis for the articles arose from investigations of three researchers: Dr. Willy Lange, University of Cincinnati; Dr. L. E. Kuentzel,

FIGURE 7.1: *The cover of an issue of Canadian Research and Development devoted to showing that carbon, not phosphorus, was the element causing eutrophication.* Printed with permission from Rogers Publishing Ltd.

Wyandotte Chemicals Corporation; and Dr. Pat C. Kerr, U.S. Environmental Protection Agency.[5] All three claimed carbon dioxide was the major factor limiting the production of algal blooms.

The introductory sentence of the articles by Legge and Dingeldein on the Lange–Kuentzel–Kerr thesis illustrates the aggressive tone that reminds one of the rhetoric used today by climate warming skeptics[28]:

> *Has the International Joint Commission, in one fell swoop, tried to discredit the research and testimony of reputable scientists, apparently twisted government science to political ends, apparently ignored gross violations of the Treaty of January 11, 1909, to which it owes its existence, and been a party to what may prove to have been the most incredible scientific/political hoax in the history of Canadian and American relations?*

An overnight best-seller in the detergent world, reprints were ordered by the industry and widely distributed among industrial magnates and governmental pollution control authorities. Within weeks, copies appeared in every country of the western world where the detergent issue was at stake, even some behind the Iron Curtain.

Another bizarre piece of propaganda was brought into play. Most limnologists of the day pointed to Edmondson's data for the recovery of

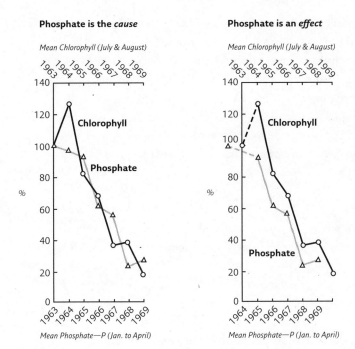

**Phosphate is the *cause***

Mean Chlorophyll (July & August)

Chlorophyll

Phosphate

%

Mean Phosphate—P (Jan. to April)

**Phosphate is an *effect***

Mean Chlorophyll (July & August)

Chlorophyll

Phosphate

%

Mean Phosphate—P (Jan. to April)

FIGURE 7.2: *A figure from a U.S. Soap and Detergent Association propaganda pamphlet. It purported to show that the decline in algal blooms in Lake Washington was not caused by decreased phosphorus as W.T. Edmondson had claimed (Edmondson's figure on the left). They claimed that declining phosphorus was the result of declining algae, by mysteriously modifying Edmondson's figure as shown on the right.*

Lake Washington as key evidence for the effectiveness of controlling phosphorus. Edmondson had pointed out that the decline in algae following sewage diversion was in direct proportion to the decline in phosphorus. In a strange twist of logic, the detergent manufacturers published their own interpretation: that the decline in algae had caused the decline in phosphorus (Figure 7.2)! As Dr. Joe Shapiro exclaimed when shown their interpretation: "But that's about as logical as claiming that lung cancer causes cigarettes!"

As described in Chapter 9, the roles of carbon and phosphorus in eutrophication were the subject of whole-lake tests at the Experimental Lakes Area, which helped to set managers on the correct course of action.

Government officials were concerned that the phosphate issue might be shelved indefinitely, due to public confusion over what appeared to be (but was not) a scientific controversy.

After initial governmental fears subsided over possible reactions from an uninformed public, matters returned to their former state. The wild and emotion-packed claims slowly dwindled into insignificance. Toward the end of 1970, most manufacturers seemed prepared to go along with what was then governmental policy of both Canada and the United States—to remove phosphates from detergents. Before examining in detail the position adopted by the U.S. Soap and Detergent Association in the phosphate–eutrophication controversy, the meaning of the word *detergent* as it is commonly used needs explanation.

## Detergents

IN EARLIER TIMES, cleaning was usually done with soaps, formed by reacting sodium or potassium salts with strong bases, such as sodium or potassium hydroxides. Performance for washing clothes was highly variable, depending on the chemistry of water in which they were used. In the mid-20th century, synthetic detergents were formulated which had more consistent cleaning properties for laundry. A detergent is a cleansing product in which the one essential ingredient is a *surfactant,* the true cleansing agent. In addition to surfactants, detergents may contain *"builders,"* ingredients with little or no cleansing power that make the surfactant work better by complexing the calcium and magnesium ions involved in water hardness; fillers, ingredients that reduce overall manu-facturing cost or act in other seemingly mysterious ways; and a variety of other substances such as artificial brighteners and perfumes that make the product seem good.

In most synthetic detergents used in the 1960s, the surfactant was the sulfate of a fatty alcohol, synthesized from petroleum products. (In soaps, the surfactant is the sodium salt of a fatty acid, derived from animal tallow or vegetable oils.) The surfactant disperses particles and stains on dirty surfaces, causing them to be suspended and removed in the wash water. In solid detergents designed for use in automatic laundry and dishwashing machines, the surfactant is a relatively minor component, typically forming only 15–20% of the weight of the product. Much of the remainder consists of builders, typically (in 1969) phosphate-rich ingredients that formed from 30% to 70% of the product. The most commonly used builders at the time of the phosphate–eutrophication controversy were sodium triphosphate in "heavy-duty" detergents designed for use in laundry machines, and other types of phosphates (pyrophosphates and metaphosphates) in products designed for automatic dishwashing machines.

Light-duty detergents, such as liquid detergents, were products designed for washing dishes and delicate fabrics by hand. They did not contain phosphate builders to any appreciable degree. Why not? In most instances, they weren't necessary. When articles are washed by hand, the human eye locates spots and stains, and the mechanical energy of the body scrubs them away. The problem with machines is, being blind, they are unable to apply their mechanical energy where it is needed most. As a result, additional ingredients (builders) have to be added to improve the cleansing power. There is also a technical reason for the lack of phosphates in liquid detergents. Phosphate polymers slowly break down in the presence of water, losing their effectiveness as builders.

Now that the purpose of detergent builders and their link with automation has been established, the next step is to examine how well the role of builder was filled by the polymeric phosphate compounds used in 1969. In the presentation the U.S. Soap and Detergent Association made to the Reuss Committee and IJC, seven functions of phosphates in detergents were listed. Phosphates were said to (i) increase efficiency of the surfactant, (ii) keep dirt particles in suspension, (iii) furnish necessary alkalinity for proper cleaning, (iv) buffer water against changes in alkalinity, (v) emulsify oily and greasy soils, (vi) soften water by tying up objectionable minerals (iron, calcium, and magnesium), and (vii) reduce levels of germs on clothes.

Given such a list, it seems inconceivable that any other single ingredient could be found to fulfill all seven essential functions. On closer inspection, however, the functions turn out to be three: (i) to soften water (the first and sixth functions); (ii) to create and maintain high alkalinity (the third and fourth functions); and (iii) to remove dirt particles (the second and fifth functions).

The role of detergent phosphates in reducing bacteria on clothes was perhaps primarily mentioned for its psychological effect. If detergent phosphates had been that essential for the elimination of bacteria, they would have found unique application in medicine, rather than in washing machines.

The prime function of detergent phosphates among the three listed is in softening water, thereby preventing calcium and magnesium ions from forming inactive complexes with surfactants and insoluble precipitates with soaps. *Detergent phosphates, in other words, were primarily disposable water softeners.* There is nothing unique about them, either in this regard or in the creation and maintenance of high alkalinity. Detergent phosphates do, however, have a special effect in suspending particles of dirt and soil, but the high concentrations used in 1969 were quite unnecessary for that purpose.

For those who might ask why a water softener has to be disposable, the answer is that it doesn't. An ordinary water softener (or, for that matter, soft water) in the supply line to a washing machine can achieve the same result as any disposable water softener. A major part of the problem of detergent phosphates is that manufacturers understandably want to make sure their product works well under the worst conditions. Therefore, they added far more builder than is necessary in most areas, in order to guarantee good performance in areas of very hard water. This matter will be returned to later, when the balance between clean clothes and clean lakes is examined. Before that, however, some of the environmental arguments used by the detergent industry deserve a "hearing" on their own.

## The Environmental Arguments

SEVERAL ENVIRONMENTAL ARGUMENTS were brought forward by the U.S. Soap and Detergent Association and the scientists who supported their stand at hearings of the Reuss Committee and the IJC. These tended to minimize the importance of phosphate (particularly detergent phosphates) as a factor in the cause and control of eutrophication. A detailed examination of these arguments leads one to conclude that, at the time of the hearings, one or both of two things must have been true: detergent manufacturers were either incredibly ignorant of the environmental impact of their products, or they realized the impact but decided to ignore it. To bring this out, a few of the more prominent environmental arguments are cited, together with errors and a statement of rebuttal.

**Argument I** "There is no conclusive proof at this time that detergent phosphates are the key element in accelerated cultural eutrophication. Indeed, there is a growing body of facts in scientific circles that suggests that nutrients other than phosphates or combinations of nutrients may be much more critical and perhaps more easily controlled in triggering this water problem … To single out only one nutrient, such as phosphate, is unrealistic in attempting to control the accelerated growth of algae and other symptoms of eutrophication." (Quoted from the statement of the U.S. Soap and Detergent Association at the Reuss Committee and IJC hearings.)

*Error* The real problem was not the cause of man-made eutrophication, but how to bring man-made eutrophication under control. Because heavy-duty laundry detergents in 1970 contained phosphates, they were implicated as contributing factors to both cause and control. The Association not only confused cause with cure but created the erroneous impression that there is something unique about detergent phosphates in regard to eutrophication, which there is not. They added further chaos by magnifying, out of all proportion, exceptions to the rule. This was irresponsible in an area of major public concern.

*Rebuttal* The topic under discussion was control of eutrophication, not cause (except insofar as it pertained to control). Among the various nutrient elements required by aquatic plants, only phosphorus is easily controllable by humans to an extent that can reduce the incidence of algal blooms. By reducing man-made supplies of phosphorus to lakes, phosphorus can be made growth-controlling, regardless of whether it was or wasn't before sources of the element are controlled. Finally, the removal of phosphate from sewage has the same effect on algal growth in most receiving waters as the removal of all nutrients combined.

**Argument 2** "Complete elimination of phosphorus from detergents, sewage, and industrial wastes discharged into Lake Erie would still result in average phosphorus concentrations of 40 micrograms per litre, four times the accepted level for algal control." (Quoted from the statement of the U.S. Soap and Detergent Association at the Reuss Committee and IJC hearings.)

*Error* Whoever made this calculation neglected to take into account the fact that 85% of phosphorus entering Lake Erie each year is removed from the water by natural processes, and deposited as sediment. Also, the "commonly accepted level" of phosphorus for algal control at the time was 10 micrograms of *inorganic* phosphorus per litre. The corresponding level of *total* phosphorus (more pertinent to this calculation) is two to three times that amount.

*Rebuttal* Taking both errors into consideration, the calculated level of inorganic phosphorus in Lake Erie under the stated conditions should have been 3 micrograms per litre, well below the accepted level for algal control. Note from DWS—early in the 21st century, these average values are approached, although temporal variation is considerable (Charleton and Milne 2004).

**Argument 3** Phosphates released from sediments in shallow lakes will stimulate algal growth even after removal of phosphates from detergents and at sewage treatment plants. (JRV's summary of a statement presented by the U.S. Soap and Detergent Association at the Reuss Committee hearings.)

*Error* This was a tactical error because it suggested the importance of phosphate to algal growth. More significantly, it showed a lack of understanding of the behaviour of phosphorus in lakes.

*Rebuttal* The *net* flow of phosphorus is from water to sediment, rather than the other way around. If, however, the release of phosphorus from sediments is as important as claimed, it adds more weight to the immediate necessity of a program for phosphorus control. The logic is simple—the more phosphorus that goes into a lake, the more stored in sediments available for release at a later time, and the longer it will take to control man-made eutrophication once it has started.

**Argument 4** The addition of nutrients to samples of water from the western end of Lake Erie shows that phosphorus is neither the sole nor the main factor limiting algal growth. *Nitrogen limits algal growth more than phosphorus.* Therefore, the removal of phosphates from detergents and at sewage treatment plants will not solve eutrophication problems in Lake Erie, because phosphate is not the main controlling factor. (JRV's summary of the statement of Dr. Willy Lange at the hearings of the IJC).

*Error* The cause of eutrophication (nutrient enrichment) is confused with the suggested cure (phosphate depletion). Also, the fact that nitrogen limits algal growth more than phosphorus in the western end of Lake Erie can be interpreted as evidence of phosphate pollution.

*Rebuttal* Sewage effluents contain high concentrations of phosphorus relative to other elements required for the growth of aquatic plants. As a result, one of the typical responses of unpolluted lakes to the addition of sewage effluents is the change from a state of phosphorus deficiency to a state of nitrogen deficiency. This is because of the high ratio of inorganic phosphorus to inorganic nitrogen in sewage (1P:4N or less by weight) relative to the needs of plants (1P:7N by weight) and the supply from natural sources (1P:14N or greater by weight). The fact that, in 1969, algal growth in the western end of Lake Erie was more limited by nitrogen than phosphorus only showed that the area was heavily polluted with phosphorus, probably from detergents.[6] The elimination of man-made supplies of phosphorus may not solve all of the eutrophication problems of Lake Erie, but it will certainly reduce those brought about by man.

Other arguments could be used to illustrate the unfortunate position of the detergent industry in greater detail; however, the four cited above suffice to make the point. The industry was environmentally illiterate at the time.

## Clean Shirts or Clean Lakes?

It remained to ask how a better balance could be achieved between clean shirts and clean lakes. There have been a number of similar environmental controversies since World War II where the scales have been tipped sufficiently to merit discontinuation of a widespread practice. Examples include the elimination of atomic explosions in the atmosphere; reducing the use of mercury, DDT, and other toxic chemicals in agriculture; and the reduction of acidifying emissions from power plants and smelters. However, there was one case that was well known to the manufacturers of detergents, in which the scales had been tipped and readjusted with considerable success. This was in relation to biodegradability of surfactants in detergents.

When synthetic surfactants were first introduced on a large scale at the close of World War II, they were greeted with enthusiasm. They cleaned dishes and clothes efficiently and without formation of insoluble precipitates and greasy scums, as had been the case with soaps. During the 1950s, however, environmental problems appeared. In the vicinity of sewage treatment plants and in waters downstream from major communities, unsightly and persistent foams accumulated wherever water was churned. In some areas, dirty, brown-coloured foams blew for kilometres around the countryside on windy days, creating health hazards and unpleasant sights.

Under U.S. governmental pressure, the detergent industry solved the problem. A relatively minor change was made in the chemical structure of synthetic surfactants that made them more susceptible to biological degradation. The foams disappeared. Detergency remained unimpaired. The entire ecosystem benefitted.

The phosphate problem, on the other hand, was not one that biodegradation could solve. The polyphosphate compounds used in detergents readily break down in water to form inorganic phosphate ions, which are then actively taken up by organisms, but not broken down.

. The possible solutions to man-made eutrophication that could be achieved through treatment of sewage to remove phosphate were examined in Chapter 6. The amounts of phosphorus used in detergent formulations and how they might be reduced are treated as a separate question, for the practical and economic reasons identified by the IJC in its final report.

Three important components of the cleansing process are often overlooked in our obsession with detergents as an essential feature of modern life. These are water, mechanical energy, and high alkalinity (pH). Water acts by its solvent effects, mechanical energy by loosening dirt and exposing the surface to be cleaned, and alkalinity by loosening and suspending particles of dirt. On the banks of the Ganges, many women wash clothes of the well-to-do without any soap or detergent at all. They use washing soda (sodium carbonate, an alkaline salt), water, and muscle power. The clothes come out spotlessly clean.

The importance of water as a cleaning agent was neatly brought out during the hearings of the Reuss Committee by citing the results of tests conducted at Marquette University in the United States. These tests involved a comparison of the cleaning power of (i) Heinecke's "no-phosphate B-5"; (ii) Procter & Gamble's Tide; and (iii) water alone. Here is the story as it was presented:[7]

Four dozen new white cotton diapers were employed for the test. All except three diapers were soiled thoroughly with water containing garden soil, and dried. Just prior to washing, the corners of all the diapers were stained with a barbecue sauce and randomly divided into three piles of 14 diapers each. Three soiled diapers were set aside for comparable purposes. Each load of diapers was washed separately in a recent model automatic clothes washer using a hot water wash for 14 minutes and warm rinse. One load was washed in plain water (Milwaukee water hardness is about 120 milligrams per litre as $CaCO_3$), a second was washed with one cup of Tide (77 grams), and the third was washed with the same weight of B-5 (77 grams). After the washing cycle, the clothes were dried for 40 minutes in a gas dryer. The three loads were compared with the original new diapers, the soiled diapers and with each other... .

Admittedly it is difficult to come to any final conclusion on detergency on the basis of the limited data available. A visual comparison was made of the three loads in comparison to new diapers and the soiled diapers. Below are presented the relative results of six observers (three men and three women) under daylight, incandescent, and fluorescent lighting conditions. The lowest number indicates the best detergency conditions.

|     |                                                      | Water | Tide | B-5 |
| --- | ---------------------------------------------------- | ----- | ---- | --- |
| I.  | Whiteness (none matched whiteness of new fabric)     | 2     | 1    | 3   |
| II. | General cleanliness (removal of soiling)             | 1     | 1    | 1   |
| III.| Stain removal                                        | 1     | 3    | 2   |

Quite different results would probably have been obtained, of course, had the tests involved grease spots or soils on cuffs or collars of shirts. Nevertheless, the results illustrate superbly the importance of water as a cleansing agent.

To the three factors of water, mechanical energy, and alkalinity, a surfactant is added that spreads out and emulsifies greasy and oily materials. Last is the builder, which is there primarily to complex ions that interfere with the efficiency of the surfactant. Looking at it this way, the detergent phosphate problem didn't seem hard to solve. Here are some of the possible solutions.

1.  Do away with automatic machines, because they are the only places where phosphate builders are extensively used (an unnecessary step that would create as many or more problems than it solved).
2.  Shift over to European-type washing machines that operate at higher temperatures, use less water, and use less detergent. (In fact, manufacturers are now doing this, in the move to front-loading machines).
3.  Revert to soap. (Only a few strict environmentalists and some mothers with children in diapers are in favour of that on a large scale.)
4.  Package builders separately from the remaining components, adjusting the amount of builder to the hardness of water in each home.
5.  In hard-water areas, use an ordinary water softener for water leading to automatic washing machines. (In fact, many homes now soften all their water.)
6.  Use a liquid detergent in conjunction with washing soda for laundry operations, automatic or not. (Dr. Alan Prince, Director General of Environment Canada's Inland Waters Directorate in 1970, said his wife had been doing this for years.)
7.  Reduce phosphate content of heavy-duty laundry detergents to less than 5% expressed as $P_2O_5$. This would have only a negligible effect in homes supplied with soft water, although it would reduce cleansing efficiency in hard-water areas.
8.  Develop replacement builders that are more friendly to the environment.[8]
9.  Be a little dirtier with a more "human" smell (not appealing to many members of society).

When the detergent industry addressed itself to the problem, only the eighth solution was seriously considered as a solution. Although several hundred replacements for sodium triphosphate had been tested, only a few showed promise in terms of all requisites to be fulfilled—cost, fabric safety,

enhancement of cleansing power, machine safety, environmental acceptability, and lack of hazards to human health.

Those familiar with a variety of today's environmental problems will
probably recognize the tactics used by the soap and detergent companies
as described above. Such tactics have been widely used by industries with
vested interests to obfuscate concerted actions on environmental issues
such as greenhouse gas emissions, acid rain, clear-cut logging, and many
other issues. One familiar tactic is to find a few scientists (there always are
a few) who disagree with the mainstream view and use their work to publi-
cize the implication that there is widespread controversy, using the popular
media rather than the scientific press to do so. A second tactic is to use the
minority studies to imply that much more science must be done before an
intelligent decision can be made, then hand over a few dollars to selected
scientists to study the problem, delaying a decision by several years.

In 1970 a compound called NTA, sodium nitrilotriacetate, was the most
prominent of all the contenders. Its history is discussed next.

# 8 | The Year of NTA

*A description of the hazards of NTA, a proposed replacement for phosphates in detergents in the ealy 1970s, in the context of the detergent phosphate controversy in North America*

## Complexing Agents

IT IS OFTEN HALF-JOKINGLY SAID in scientific circles that, for every chemical compound headlined by science or industry today, there is always some German chemist who first isolated or synthesized it in the 19th century. Nitrilotriacetic acid (NTA) is not an exception to this rule. W. Heintz of Germany first described its isolation and synthesis.

Heintz was studying the reaction products formed when ammonia and chloroacetic acid were mixed together. He identified four products, all derivatives of acetic acid—hydroxyacetic acid, aminoacetic acid, iminodi-acetic acid, and NTA. So far as is known, Heintz gave no thought to the possible industrial significance of any of these compounds nor did any other chemist for a period of 73 years.

In 1935, Dr. F. Muntz, working at the I.G. Farbenindustrie plant in Mainkur, Germany, began examining the possibility that aminopoly-carboxylic acids, such as NTA, might be of use in the dyeing and textile industries. When Muntz repeated Heintz's study of the reaction between ammonia and chloroacetic acid he noticed an unusual property, one that turned out to be of considerable importance to the commercial interests of I.G. Farbenindustrie. Muntz observed that something in the reaction mixture interfered with the precipitation of calcium and magnesium salts.

The significance of this observation lay in the fact that doubly and triply charged ions, such as calcium, magnesium, iron, and aluminum, are a common cause of problems in dyeing processes due to formation of insoluble precipitates with dyes. If these or other metallic ions reach appreciable concentrations in dye baths, uneven and unpredictable colouration of fabrics can result. Following the general rule that an ounce of prevention is worth a pound of cure (as true in the chemical industry as it is in medicine), Farbenindustrie was actively searching for materials that could prevent the formation of precipitates. NTA was such a chemical.

When Muntz discovered that the principal calcium-binding compound in the reaction mixture was NTA, Farbenindustrie immediately took out patents for its use in the dye and textile industry. Production was initiated in 1936, and the first lots were distributed under the trade name Trilon A in 1937.

Another compound similar in general structure to NTA was also marketed by Farbenindustrie at about the same time. Originally called Trilon B, it is now more commonly referred to as EDTA (ethylenediaminetetraacetic acid). As a result of its lower production cost and greater complexing strength, EDTA rapidly eclipsed NTA in terms of scientific and industrial use.

NTA and EDTA are members of a class of chemicals known as aminopolycarboxylic acids. Similar to amino acids that make up body proteins, the principal significance of these compounds lies in their ability to form water-soluble complexes with metal ions. The structure of the NTA-metal complex resembles that of a human hand enclosing a tennis ball—the thumb corresponding to the amino group of NTA, three of the fingers to the carboxyl groups of NTA, and the tennis ball to a metal ion.

The use of EDTA in the dyeing and textile industry was followed by its application in electroplating, developed by H. Brintzinger in Germany during the early part of World War II. Shortly after this, Gerold Schwartzenbach of Zürich began a detailed series of investigations of the structure and properties of aminopolycarboxylic acids. Since that time, EDTA has found a wide variety of applications—in water softening and removal of boiler scale; in cosmetic and pharmaceutical industries; in manufacture of rubber, polymers, and plastics; in the leather industry; in photographic processes; as a food preservative; in plant nutrition studies; as a food additive; and in beverages. The principal role played by EDTA in these various applications is the complexing of metal ions, thereby preventing formation of unwanted precipitates or oxidation reactions catalyzed by metal ions.

In contrast to EDTA, NTA remained as a "specialty" chemical because of its comparatively high cost per unit of complexing power. This changed in 1959 when the U.S.-based Hampshire Chemical Corporation, later to become a division of W.R. Grace & Co., developed a new process that not only reduced the manufacturing cost of NTA but also increased the purity of the product. The first commercial use of NTA was as a water-softening and water-conditioning agent.

When the capability for U.S. production of NTA increased in 1966 with the expansion of Hampshire Chemical's facilities, wide-spread consideration was given to other possible uses—among them, the possibility that NTA might replace sodium triphosphate (STP) in detergents. NTA was more expensive than STP, roughly 12–14 cents versus about 8 cents per pound (U.S. prices in the 1960s) in large industrial lots, but it was also 1.4–1.8 times more effective than STP in water softening per unit of weight. The cost differential still favoured STP as costs are calculated by industry. The true comparison was unknown in terms of overall ecosystem costs—the total costs, taking all human and environmental matters into consideration.

One of the first companies to scrutinize NTA as a detergent "builder" was Procter & Gamble. This was not accidental. It illustrates a general evolutionary rule, as true in the competitive world of industry as it is in the competitive world of biology, that as soon as a new product meets with success on the market, work begins to develop newer and better substitutes.

Procter & Gamble was responsible for the introduction of Tide in 1947. It was a product that received a phenomenal reception from the buying public. The secret of Tide lay in the combination of a synthetic surfactant with STP. Both these chemicals were being produced industrially in the 1930s, although for separate purposes. Procter & Gamble chemists discovered that when the two were put together a super-detergent product was created. When Tide was being developed, however, STP was not available in the quantities required. New manufacturing facilities had to be constructed and extended purchasing commitments made before it became a commercial reality. Building on the tremendous success of Tide, Procter & Gamble continued to search for new and improved formulations.

By the time of the phosphate controversy in 1969–1970, Procter & Gamble had assessed several hundred compounds as potential replacement builders for STP. Each of these, alone and in combination with other detergent ingredients, had to pass through screening tests designed to meet four primary criteria—effectiveness in enhancing cleansing action; safety to fabrics and washing machines; safety in terms of household use and human health; and environmental acceptability. While none of the compounds tested showed the same across-the-board performance as STP, there was one that looked exceedingly good as a partial replacement. It was NTA.

From information presented by Procter & Gamble to the Subcommittee on Air and Water Pollution, chaired by Senator Edmund Muskie in the spring of 1970, it emerged that the company had been working on NTA since 1961, one year before the Hampshire patent was issued. Procter & Gamble further told the Muskie Subcommittee that between 1965 and 1969 more than $11,000,000 had been spent on development and testing of replacements for STP. The estimated budget for 1970 was $3,500,000.

In spite of the fact that Procter & Gamble, more than any other company, had the largest body of information on all aspects of NTA, it was not the first company to market a detergent product containing the compound. A.B. Helios in Sweden began marketing detergents containing NTA as a partial replacement for STP in 1967.

The extent to which A.B. Helios tested the environmental and human health safety of NTA is not entirely clear, although it was far less than Procter & Gamble had done. As a result of the happy combination of a good sales promotion, effective products, and a phosphate-conscious public, NTA-containing detergents had captured 15% of the laundry detergent market in Sweden by 1970.

Why did Procter & Gamble delay doing anything other than test-marketing two products slightly enriched with NTA? One reason was the stickiness of the trisodium salt of NTA. Even when mixed and diluted by other ingredients it tended to collect moisture, becoming stickier than salt in a shaker on a humid day.[1] The problem was not insoluble, but it did create difficulties in both manufacturing and packaging. Another reason Procter & Gamble held off—hard as it may be for environmentalists to understand—was that the company did not feel there was sufficient environmental and human safety information about NTA at that time. Knowing more than any other company was evidently not enough. The reason may have been that, because Procter & Gamble is so visible (as is the case with all large companies), it cannot be irresponsible and survive.

## The Year of NTA

THE YEAR 1970 will go down in the annals of detergent history as the year of NTA. NTA burst into prominence in February. After several months of serious consideration, interest by U.S. officials dropped in December. The case is of particular interest, because NTA had been studied more than most synthetic chemicals for its effects on human health and the environment. An analysis of the way it was viewed by government, industry, and the public in 1970 shows how drastically views on the environment had changed since World War II.

The year began with the hearings of the 1970 International Joint Commission (IJC) on water pollution in lakes Erie and Ontario in late January and early February. Those present at the six hearings saw little need to await the final report of the IJC in regard to the recommendation on phosphates in detergents. From statements made by representatives of the two governments at the hearings, it looked as though Canada, the United States, and the IJC would probably all push for a zero level of phosphate in detergents "soon," if not by the end of 1972. Canada announced it would introduce preliminary regulations on the phosphate content of detergents within 6 months.

Even before the representatives of government and industry departed for their homes, steps were being taken at their head offices to prepare the ground for the next strategic moves. Recognizing that the rules had completely changed, a number of detergent manufacturers rolled up their sleeves to meet the new situation. Among the various replacements discussed were NTA, citric acid, polycarboxylates, carbonates, and silicates. Assuming that NTA would become a "chemical of consequence," government workers immediately planned a broad assessment of the environmental effects of NTA.

Aware of variations in composition of different batches of chemicals produced on a large scale, the manufacturing industry made available a ton of NTA from several sources for various scientific tests. The sample was thoroughly mixed and subpackaged by a U.S. government agency for distribution to various other agencies in Canada, Sweden, and the United States.

Government plans called for a diversified array of environmental tests. Among these were the effects of NTA on fish and other forms of aquatic life, photosynthesis and reproduction of algae, release of heavy metals from sediments, sewage treatment processes, integrated behaviour of freshwater ecosystems, movements of metallic ions in groundwater, soil systems, and last, but not least, human health.

In addition, studies were made of the breakdown rate of NTA under various natural conditions, adaptation times of microorganisms prior to metabolizing NTA, susceptibility of various NTA–heavy metal complexes to biological breakdown, decomposition, products and prediction of equilibrium concentrations of NTA and NTA–heavy metal complexes under various conditions.

One of the more amusing, although frustrating, incidents during these investigations was the complete destruction by the resident muskrat (*Ondatra zibethicus*) population of a long-term experiment involving 20 plastic enclosures suspended from a frame on floats in Clear Lake, Manitoba. Dr. Robert D. Hamilton, in charge of the Fisheries Research Board of Canada's (now the Canadian Department of Fisheries and Oceans) work on NTA, couldn't quite believe it when he heard the news. Burrowing into the floats of the platform supporting the tubes, a family of muskrats had found styrofoam to be a modern home-building material very much to their taste. Easy to chew and with excellent insulation qualities, the muskrats were delighted with the improved living conditions Dr. Hamilton provided.

When the research team arrived one day to take their weekly samples from the tubes, they found everything in chaos. Some muskrats had fallen into the open tubes, scratching and biting through the thin plastic in order to escape. All the plastic enclosures were ruptured. To complete the farce, the muskrats chewed off the four guy lines that anchored the float, setting the raft adrift on the lake. As Hamilton dejectedly viewed the mess, he couldn't refrain from thinking what an excellent coat the muskrats would have made, valued at about $50,000 in terms of the damage done.

One of the most serious deficiencies in regard to NTA at the beginning of 1970 was the lack of a sensitive, specific, and accurate method of analysis for use in natural environments. The only standard technique available was based on the complexing power of NTA for metal ions. The problem was that in natural soils, sediments, and water, there are many natural complexing agents. Under ordinary conditions of analysis with the standard technique, these were indistinguishable from NTA.

Fortunately, before most of the studies were under way, Dr. Peter Goulden and Dr. B.K. Afghan of Environment Canada's Water Quality Branch developed a specific and sensitive polarographic method for analysis of NTA. At the same time, Dr. Domenico Povoledo of the Freshwater Institute in Winnipeg developed a new technique based on mass spectrographic analysis. Within a few months, several other methods based on gas chromatography of NTA derivatives were also available. Without these methods, many questions regarding NTA would have remained unanswered at a critical time. At the start of 1970, the standard method was nonspecific and only useful for concentrations of NTA above 0.2 parts per million (ppm) of water. At the close of 1970, several specific methods had been developed permitting specific detection of 0.01–0.02 ppm of water. Methods are now available that can detect as little as 0.001 parts per billion of NTA in water.

As tests to determine environmental acceptability of NTA were being developed by scientists working in various government agencies, industries, and universities, new products started appearing on supermarket shelves with "no phosphate" labels. What they contained was almost anyone's guess, for there were no legal requirements for labeling at the time in either Canada or the United States. Some contained NTA, while in others, borax and washing soda (old-fashioned chemicals for softening water) took the place of STP. On chemical analysis, one product turned out to contain 50% common salt, an ingredient that could only have been used to make the weight of the box seem good for the price.

## The Race is On

THE REPORTS of the Reuss Committee and the IJC appeared in early spring 1970; both recommended the elimination of phosphates from detergents by 1972. Procter & Gamble beat them to the gun with a full-page advertisement in a number of prominent U.S. and Canadian newspapers stating they were on the team. Without admitting any proven connection between detergent phosphates and man-made eutrophication, the advertisement said that Procter & Gamble was prepared to cooperate just on the possibility such a connection might exist. A Procter & Gamble product would soon appear on the market with a 25% replacement of STP by NTA.

Other manufacturers likewise came out with new products and packages. Companies with a small cut of the detergent market made grandiose claims for their products, whereas large companies often apologized for being forced to market products that, while good, were said to be inferior to their previous products. Detergent products began to appear from manufacturers that no one had heard of before. There were persistent rumors that the Mafia had entered the field. For a while, it seemed as though cleanliness and even sex appeal had been replaced by environment appeal—witness the names of products such as Concern, Ecolo-G, and Un-Polluter. New STP substitutes, including NTA, were slowly entering the environment.

As the year of NTA advanced, reasons for the importance attached to NTA as a replacement builder became apparent. In testimony before the "Muskie" Subcommittee on Air and Water Pollution (of the Senate Public Works Committee) in May, William. C. Krumrei of Procter & Gamble revealed the extensive testing of NTA the company had performed in regard to human and environmental safety. Because of possible interactions with other compounds, many tests had to be performed not only with the trisodium salt of NTA (the probable form of NTA in detergent formulations) but with NTA as part of specific formulations likely to be used. The list of human safety tests performed by Procter & Gamble, reproduced below from testimony presented by William Krumrei, reveals some of the many questions that had to be answered by a responsible company, before a product is marketed in the modern world. The Chronic Feeding Study involved three levels of NTA in diets fed to rats over a period of 2 years. Headings discussed by Krumrei in relation to human safety were as follows:

> *Routine-Type Animal Studies (acute oral toxicity, emetic activity, eye irritation, percutaneous toxicity); Human Skin Studies (sensitization tests, laboratory skin tests, clinical mildness tests, diaper clinical test); Metabolism Studies; Reproduction and Embryogenic Studies; Sub-Acute Feeding Studies (mineral excretion, manifestations of toxicity); Chronic Feeding Study; Estimation of Maximum Level of Human Ingestion of NTA from Detergents (residue from dishes, pots and pans, residues in tap water); Determination of Factors of Safety (chronic toxicity, teratogenesis); Safety as Regards NTA Contribution to Nitrate Levels; Inhalation Studies; Carcinogenicity Testing; Mutagenicity Testing; Enzyme Studies.*

In addition to the above, Procter & Gamble had done sufficient testing to be assured that NTA was biodegradable in sewage treatment and the natural environment and that there was no danger to human health or the environment at a level of 25% replacement of STP. In the text of the prepared statement released in May 1970, the company indicated that, in order to meet its objective of a 25% replacement, orders had been placed for $167,000,000 of NTA. The company had also committed $6,800,000 as capital expenditure to modify existing facilities. Procter & Gamble was going all the way.

Dr. Samuel S. Epstein of the Children's Cancer Research Foundation, Inc., and the Harvard Medical School also presented testimony before the Muskie Subcommittee in May 1970. Dr. Epstein raised many unanswered questions about NTA. One was the possible significance of the decomposition of NTA in increasing nitrate levels in water. (Nitrates and nitrites are implicated in methemoglobinemia, a condition of reduced oxygen transporting power of the blood due to inactivation of hemoglobin by nitrite. Infants are especially sensitive to this condition.) He also raised the specter that NTA transformation products might be carcinogens (cancer-causing agents), teratogens (agents causing birth defects), or mutagens (agents causing mutations in cell nuclei). Finally, would use of NTA in detergents create problems in redistributing heavy metals or enhancing their toxicity to humans?

In contrast to this last suggestion, one of the earliest proposed uses of NTA in the environment was to reduce the toxicity of metals such as zinc and copper to fish. Dr. John Sprague, then at St. Andrews Biological Station of the Fisheries Research Board of Canada, had shown that NTA protected Atlantic salmon (*Salmo salar*) and other fish from toxic effects of copper and zinc released from base-metal mining activities.[2] From a fish's point of view, NTA apparently looked good, but what was the case for humans?

## Resolving the Controversy

TOWARD THE END OF THE SUMMER of 1970, plans were developing for a closed meeting of government scientists from Canada, Sweden, and the United States to assess the environmental acceptability of NTA. The meeting was scheduled for the Canada Centre for Inland Waters in Burlington, Ontario, December 8–10, 1970.

As delegates assembled the night before the meeting, it was apparent something unusual was in the wind. The next day, Dale Chernoff and Diane Courtney, two scientists from the U.S. National Institute of Environmental Health Sciences (NIEHS), presented results interpreted as showing that NTA had enhanced the toxicity and teratogenicity (production of birth defects) of cadmium and methyl mercury administered to rats and mice.

The primary question was whether the experiments were pertinent to normal household use of NTA-containing detergents. The cadmium experiments were performed by injecting solutions into experimental animals. The concentrations of chemicals used were high. More problematic was that some of the supposedly inert components of the test materials were varied in an unsystematic way, invalidating certain conclusions. The investigators themselves stated that they did not know how to place the results in the perspective of real life. In short, the experiments were poorly designed, with no interpretation of how the results pertained to the projected use of NTA.

Following the presentation of results from other NTA experiments, none of which was damning, the general feeling on the part of scientists present was that NTA would hang under a heavy cloud of suspicion until the NIEHS report could be properly interpreted. The results had to be released because of their possible importance. However, without time for documentation and perspective, what official stand could be taken but against NTA? Having taken such a stand, could NTA ever recover, even if it proved to be lily-white pure? For all practical purposes, it looked as if NTA was as good as dead, perhaps more for psychological reasons than for considerations based on environmental or human health.

In terms of a zero-risk policy there was only one thing to do—stop the use of NTA until the situation with respect to human health could be more clearly defined. At least, that was the U.S. government view. As will be seen later, Canada and Sweden took a different stand.

Events moved rapidly in Washington after the December 8–10 meetings in Burlington. Following high-level discussions, a series of calls went out to manufacturers connected with the soap and detergent industry on Wednesday, December 16. The message simply stated that corporate officials of the various companies might wish to attend a meeting pertaining to an urgent decision on NTA at 4:00 pm, Thursday, December 17.

At the appointed hour, the U.S. Surgeon General and the Environmental Protection Agency (EPA) Administrator told the Industrial executives the Chernoff–Courtney findings. Without any time to digest the findings, the big three detergent manufacturers agreed to cease production of NTA-containing detergent products. The executives were disturbed by the knowledge that U.S. government officials had discussed the findings with scientists from other countries before revealing them to U.S. manufacturers.

On December 18, 1970, a joint statement was issued by the Surgeon General and the EPA Administrator. It began, "We commend the major detergent manufacturers for their voluntary action to discontinue the use of NTA (nitrilotriacetic acid) in the manufacturing of detergents, pending further tests and review of recently completed animal studies."[3]

Repercussions from the December 18, 1970, NTA announcement were still being felt in some quarters in 1972 and continued for years. The suddenness of the shock, lightning reactions, and governmental pressure seemed incredible to those adversely affected. The ignition keys to the NTA production engines at Hampshire Chemical and Monsanto were turned off even before company representatives had returned from Washington. Detergents lacking both phosphates and NTA began an upward surge in sales. The costs to some industries were immense. It is said that Procter & Gamble wrote off $7,100,000 in binding contracts to purchase NTA.

With the U.S. federal government holding off any official regulations to limit the phosphate content of detergents, states and municipalities began to take action on their own. Regulations were enacted in New York and Indiana similar to those in Canada, and in 1972, bills were pending in many other parts of the country. Russell Train, Head of the Council on Environmental Quality, and Dr. Charles C. Edwards, Head of the Food and Drug Administration, joined Surgeon General Steinfeld and EPA Administrator Ruckelshaus for the next major announcement, issued September 15, 1971. Without any specific evidence of harmful effects and despite a refutation of the Chernoff–Courtney work, they urged that NTA "should not be used in detergents at this time because of unresolved questions concerning its possible long-term effects on health and the environment." Cautioning against increased use of detergents containing caustic materials as substitutes for phosphates, they advised the states and their political subdivisions to "reconsider policies that unduly restrict the use of phosphates in laundry detergents." The message seemed to be—phosphates in detergents are good, let's keep them and get on with removing phosphate at municipal treatment plants.

An editorial in the Toronto Globe and Mail called the statement a "shabby turn-about." U.W. Poston, Commissioner of Environmental Control for Chicago, and Jerome Kretchmer, New York City Environmental Administrator, termed the federal pressure for removal of phosphates at sewage treatment plants as tantamount to a federal order for subsidization of the detergent industry by taxpayers in Chicago and New York City. Chicago passed municipal legislation to ban phosphate detergents.

The Canadian reaction following the presentation of the Chernoff–Courtney report at the December 8–10, 1970, meetings contrasted with that of the United States.[4] The first step was to dispatch a team of experts from the departments of National Health and Welfare and Environment to the U.S. NIEHS in Chapel Hill, North Carolina, where the Chernoff–Courtney findings were discussed in detail.

Next, there was a meeting of government scientists on January 6, 1971, to discuss the scientific basis of policy recommendations to be made to the Honorable Jack Davis, Minister-designate of the new Canadian Department of the Environment. In the opinion of those present, the Chernoff–Courtney experiments were not considered pertinent to real life. The experimental design was defective, and although of interest, the results from injecting NTA in experimental animals were uninformative in terms of intended use. There was cause for a cautious approach to NTA but, in the opinion of the Canadian environmental scientists, not one that called for a ban.

What the Canadian scientists said in effect was—there is no such thing as a no-risk policy. Everything in life, even decisions not to do something, inevitably carry a risk. From environmental and human health points of view, if a fraction of a part per million of NTA in natural waters looked bad, then the occurrence of several hundred parts per million of EDTA as additives in some human foods looked worse. The group recommended that the Chernoff–Courtney experiments be repeated, extended, and interpreted in an environmental context; that possible effects of EDTA as a food additive be examined; that the NTA question be recognized; but that no recommendation on NTA pro or con be given.

After receiving this information, Mr. Davis scheduled a meeting for February 17, 1971, with leading Canadian manufacturers involved directly or indirectly in the soap and detergent industry. Stating that he intended to announce governmental policy on maximum phosphate levels in detergents in a few weeks, he asked industrialists to express their views as to what he should say—either at the meeting or by separate mail. The industrialists did both. As agreed, the Minister-designate of the Department of the Environment held the industrialists' views in confidence.

On April 8, 1971, Mr. Davis issued a news release indicating the maximum permissible levels of phosphates in detergents previously set at 20% $P_2O_5$ (8.7% as P) on August 1, 1970, would be further reduced to 5% as $P_2O_5$ (2.2% as P) on December 31, 1972. Despite pressure from segments of industry, he said nothing about NTA. After a review of new information and a further meeting with industrial representatives on May 11, 1972, Mr. Davis reaffirmed his position and the regulation went into effect January 1, 1973.

**The Year in Perspective**
WHERE DID THIS LEAVE US with respect to NTA? Unfortunately, there is no simple answer, other than to say that it is a matter of scientific evaluation by specialists and of public desire. None of the evidence (up to 1974) suggested that NTA or its metabolites would cause mutations, cancer, embryonic malformations, or enhanced toxicity of heavy metals more than STP under the same conditions. Although some deleterious effect could always turn up, NTA had been examined in far greater detail than most synthetic compounds in use. In the final analysis, the only effective test of acceptability of NTA would be through a slow buildup in use, with close monitoring along the way. This started in Sweden in 1967 and in Canada in 1970.

In a report dated April 1, 1972, Arthur D. Little, Inc., listed concentrations of NTA in samples of tap water from communities in Long Island, N.Y., and several Ontario cities. The results averaged less than 0.025 ppm (the level of detection), with some waters extending into the range of 0.025–0.10 ppm. These values were in line with predictions and within allowable limits using a safety factor of 100. (A safety factor of 100 assumes daily intake could be up to 100 times the maximal predicted amount without evidence of harm.) The report concluded that there was a very low probability of environmental or human hazard but advised continued monitoring and research to assure environmental and human safety.

Perhaps, in the end, the separate decisions of Canada and the United States were both wise. Because of its lower population size and abundance of water, Canada had more flexibility than the United States in relation to "chemicals of environmental consequence," particularly for wastes with waterborne effects. With the same per-capita production of wastes, environmental outputs were in the ratio of 1 Canada to 10 United States. In terms of NTA as a replacement for STP in 1970, it meant 50 million kilograms (110 million pounds) in Canada versus 500 million kilograms (1.1 billion pounds) in the United States. More important, should the population in both countries double between 1970 and 2010 and per-capita use remains the same, the increase in NTA for Canada would be only 50 million kilograms (110 million pounds) as compared with 500 million kilograms (1.1 billion pounds) in the United States.

A number of things pertaining to more general matters have been learned from the tête-à-tête with NTA. Those of importance are as follows:

1.  Prior to 1960, any governmental official who made adverse statements about a major industrial product on the basis of tenuous evidence would have suffered politically. In 1970, at least with regard to NTA, exactly the opposite was true. Government officials in all countries wisely feared the consequences of approving a chemical that might later turn out to be a teratogen. This conversion of "go-go" proponents of development into environmental conservatives was a healthy sign. We operated the other way when industrial growth was desired in North America; now it was time to change to a more balanced view.

2.  There is a useful place in society for persons such as Dr. Samuel S. Epstein, who persistently draw attention to possible consequences of synthetic chemicals used in high quantities. Even though none of Dr. Epstein's claims that NTA or one of its decomposition products could be a mutagen, teratogen, or carcinogen in environmental concentrations were proven correct, his challenges created an atmosphere of caution. When any chemical is studied in great detail for its total ecosystem effects, something harmful is almost certain to show up. Ultimately, we must learn to be more thorough in our analysis of new chemicals, examining risk–benefit relationships in detail after placing some maximum limits on risk.

3.  In the final analysis, the most important component of the biosphere, to man, will always be "me." That is something we may modify but cannot appreciably change. In terms of the effects of chemicals, humans had always come first, property next, and the wild environment last. In the past 40 years, as we have come to understand the importance of the natural environ-

ment in more detail, we have slowly moved toward protecting the environment with the same fervour we have always used to protect our more immediate selves.

4. In his book, *The Closing Circle*, Barry Commoner (1971) pointed out that new technological products contributed more to increase the levels of pollution between 1945 and 1970 than growth of either population or affluence. As examples, he cited the use of mercury in the chlor-alkali industry, non-returnable bottles, fertilizer nitrogen, detergents, and plastics. NTA was a chemical that falls into Commoner's "changing technology" class. In the statements to U.S. officials on NTA, the total quantities involved were said to have played an important part in the decision. Commoner's book should be required reading for high-level government officials in this time of continuing growth of new technologies and human populations.

5. There is a "demotechnic" (population multiplied by technology) growth factor that must be considered in dealing with all "chemicals of consequence" such as NTA.[5] The NTA-on and STP-off switch in the United States in 1970 involved quantities of about 500 million kilograms (roughly 1.1 billion pounds) of NTA. It pertained to 211 million people living in a fossil fuel powered, high-gear, technological state. A comparable decision in 1870 in the United States would have pertained to only 40 million people in a horse-drawn, low-gear, technological state. The difference is like making a turn in a car traveling 10 kilometres (6 miles) per hour compared with the same turn in a 20-ton truck at 100 kilometres (60 miles) per hour. Currently, we see no sign of relief. The United States now has over 300 million people. Gas-guzzling automobiles make Americans reliant on foreign countries for fossil fuels, foods, and most other commodities. Some of these foreign countries have very unstable governments. The United States has lost considerable flexibility in its decisions, not just because of technology, but because of the unabated growth of population and technology combined. Canadians who value their independence and freedom should take note, because our own reserves of inexpensive oil and gas near exhaustion, our cities continue to increase in population and industrial growth, and an increasing amount of our

resources are exported to the United States to fulfill their needs. The replacement for cheap oil and gas is to exploit the oil sands of Alberta, where bitumen is extracted from sand at enormous costs in energy, greenhouse gas emissions, and the environment. The damage to Alberta's boreal landscape is huge and irreparable. Is this truly the price that we are willing to pay for prosperity?

6. Ecosystem protocols are needed for new chemicals of environmental consequence. These protocols are descriptions of specific tests an industry must perform before governmental approval is given for a particular use. Protocols exist for drugs and food additives intended for human consumption. Now we need them for chemicals of environmental consequence. One of the lessons from NTA is that a sensitive and specific method of analysis *for use in natural environments* must be a requirement of any ecosystem protocol. Sadly, in 2005, such protocols exist for only a few of the over 100,000 man-made chemicals that are now in widespread circulation.

7. Because of the costly nature of obtaining extensive data before even preliminary approval is given to chemicals of environmental consequence ($1–10 million per chemical), some premeditated form of international cooperation is needed. Current (2007) international arrangements are improving but are still haphazard and unsystematized.

8. Through controversies such as those involving phosphates and NTA, citizens will eventually come to realize that specialists, including scientists, exist to provide information, not to make communal judgements or be responsible for public ethics. Regardless of what the facts may be the specialists' opinions and interpretations are their own. Citizens and specialists in other fields must realize that if they give their proxies to a specialist, the proxy is for a vote based on the viewpoint of a specialist.

9. Because of our preoccupation with cleanliness, efficiency, and costs, we tend to automatically assume that, when one good thing goes out, another equally good thing must come in. Perhaps through examples like NTA, governments may eventually realize this will not always be so, particularly in areas of rapid demotechnic growth. In fact, not one but several kinds of

substitutes for STP have been developed. Waste diversification for environmental stability may turn out to be just as important as product diversification for economic stability in an industrial or national sense.[6] The two, in fact, ultimately become one and the same. In the 1960s, the detergent industry was not diversified in relation to alternatives to STP. It was like a country dependent for its survival on the culture and export of a single crop.

One might think from the healthy skepticism governments had of NTA in 1970 that we had entered a new age, with a more balanced view of technological products such as STP and NTA.

Nothing could be further from the truth. NTA just surfaced as a "political" chemical. Other synthetic chemicals in everyday use remain with long-term consequences for both humans and the environment largely unknown and little studied, or known and ignored such as the effects of chlorine in swimming pools on eyes of children, of EDTA (a common food additive) on human health, of cosmetics on skin, of drugs on the mind, and of alcohol on the body.

The controversy over NTA, based on the remote possibilities of slight increases in cancer or effect on fish, is typical of the debates that rage over many chemical products and human activities today. It is an example of how spurious debates can impede progress toward rational control of waterborne and airborne pollutants. Currently, the production and dissemination of new chemicals still greatly exceeds our ability to test and regulate them.

# 9 | Understanding Eutrophication from Experiments in Small Lakes

*Some of the whole-ecosystem experiments that helped to resolve the debate over controlling eutrophication*

THE EXPERIMENTAL LAKES PROGRAM was formed in 1968. At the time the project was started, most of the scientific information on eutrophication had either been gleaned from laboratory or mesocosm size experiments or deduced from observational field studies, usually in lakes affected by a variety of human insults, making it difficult to discern cause and effect relationships. Regulators were understandably reluctant to commit millions of dollars to eutrophication control schemes for large waterbodies on the basis of such scanty evidence. Dr. W.E. Johnson, the newly appointed director of the Freshwater Institute in Winnipeg, was instrumental in convincing the now-defunct Fisheries Research Board of Canada that only experimentally polluting entire small pristine lakes with nutrients could yield unimpeachable evidence for the causes and consequences of eutrophication. Ecosystem-scale evidence could then be used to convince regulators to take the necessary action to manage eutrophication in larger lakes. Later, this concept was applied to studies of other pollutants, including sulfuric and nitric acids, radioactive materials, trace metals, reservoir flooding, and small amounts of hormones.

As mentioned briefly earlier, an area southeast of Kenora, Ontario, was selected as the site of the Experimental Lakes Area (ELA). The ELA originally included 46 pristine small lakes and their catchments. Also, local people in the area had to be convinced that deliberately polluting lakes was a sound idea!

Initial eutrophication experiments attacked the then-controversial issue of what nutrients must be controlled to manage eutrophication, many of which are described in Chapter 7. Later studies addressed the effects of water renewal, changes to food webs, and nutrient stoichiometry on eutrophication. Johnson and Vallentyne (1971) described the originally envisioned program in more detail.

### The Science of Eutrophication in the Mid-1960s

THERE WERE MANY THEORIES of what caused eutrophication in the mid-1960s, contributing directly to the confusion described in Chapter 7. Two synthesis volumes, one from a 1967 symposium published by the U.S. National Academy of Sciences (1969) and the other by the American Society of Limnology and Oceanography (Likens 1972), provided a good overview of understanding of eutrophication at the time. Individual papers in these volumes describe the contributions of many chemicals, including nutrients, major cations, and trace elements, to the eutrophication problem. Neither symposium presented a well-reasoned synthesis that would convince regulators to take any specific approach to eutrophication control.

Fortunately, the needed synthesis was being prepared elsewhere. Richard Vollenweider, then working for the Organisation for Economic Co-operation and Development (OECD), had prepared a several hundred page review, distilling decades of earlier literature. It concluded that phosphorus, followed by nitrogen, was the most likely cause of the rapid increase in eutrophication in the 20th century. Vollenweider (1968) introduced the concept of phosphorus "loading" (input) as the factor that must be considered, rather than ignoring events in the catchments of lakes, as had been the practice in earlier years. Ironically, Vollenweider attended the 1967 symposium in Madison, Wisconsin, that had produced the 1969

National Academy volume, but he was not invited to speak. However, once his comprehensive review was known, scientists and managers seized on this clearly written document and proposed it as the basis for reducing inputs of phosphorus to the St. Lawrence Great Lakes. Later, Vollenweider received the internationally renowned Tyler Prize for Environmental Achievement for his scientific work on the eutrophication problem.

## The Carbon Controversy

SOME SCIENTISTS, largely supported by the detergent industry or companies that sold phosphates to detergent makers, argued that elements other than phosphorus were responsible for the eutrophication problem. Many of their arguments were specious, and they relied much more on propaganda pamphlets than substantive publications. This lack of scientific professionalism led many scientists to refer to them in derogatory terms as "the Soapers" (a term coined by Dr. Joe Shapiro). As discussed in Chapter 7, evidence that dissolved inorganic carbon (DIC) limited algal growth, based on a few short experiments in bottles and small mesocosms (Kuentzel 1969; Kerr et al. 1970), was widely disseminated in detergent company propaganda. Small-scale experiments, using water from Lake Erie, showed convincingly that adding carbon stimulated algal growth much more than adding phosphorus did (Lange 1970).

In a predictable industry manoeuvre to delay regulation of their products, the detergent industry displayed the studies promoting carbon limitation in pamphlets, in industrial magazines, and in scientific and legislative hearings, casting doubt among decision makers that eutrophication could be controlled by simply controlling phosphorus. These shenanigans were described in more detail in Chapter 7.

The battle over phosphorus versus carbon control was heated as ELA scientists, led by DWS, began the first surveys of the ELA in 1968 and began planning the first experiment. The results of preliminary surveys revealed an excellent opportunity to test the carbon limitation hypothesis on a realistic scale. The lakes at ELA had incredibly low concentrations of dissolved inorganic carbon, much lower than the concentrations that were present in the laboratory studies and ponds where carbon had been shown to limit the growth of algae. However, the ELA lakes were also oligotrophic, with phytoplankton abundance of generally <2 micrograms per litre, measured as chlorophyll *a* concentration. DWS and Gregg Brunskill proposed to test the carbon hypothesis by adding only nitrogen and phosphorus to Lake 227, the lake with the lowest dissolved inorganic carbon concentration in our survey of the ELA. This lake was several miles from roads, and there had been no human intervention in either the lake or its catchment. The epilimnion of the lake contained only 50 micromoles per litre of dissolved inorganic carbon per litre (0.6 milligrams per litre) in midsummer. We reasoned that either the lake would remain oligotrophic, supporting the Soapers' theory (in which case we could make controlled carbon additions to examine how much carbon was necessary to promote algal blooms), or it would become eutrophic, ruling out carbon as a cause of eutrophication. Many eminent scientists of the day were called in to debate the proposal. Some were sceptical, even stating that we would further confuse the eutrophication controversy. In autumn of 1968, a late-night debate at ELA included most of the scientists in the new Eutrophication Section as well as senior managers. Richard Vollenweider came out strongly in favour of our proposal to fertilize Lake 227. This convinced others to go along with it, some grudgingly.

Although we had completed only three surveys of Lake 227 in 1968, the urgency of the question compelled us to begin the experiment quickly, relying largely on similar but unfertilized lakes in the area for reference and background information. After ice-out in 1969, Stefan Holmgren and DWS made intensive measurements of plankton species, biomass, primary production, and nutrient concentrations on Lake 227 and several similar lakes for several weeks. A V-notch weir and continuous water level recorder were installed at the outflow of Lake 227 to measure the volume of outflow. This allowed us to calculate the lake's water renewal time and nutrient losses. We began adding fertilizer to the lake on 26 June 1969. The experimental "phosphorus loading" was calculated to be about 10 times the amount that our early estimates indicated was entering the lake from natural precipitation and runoff. The N:P ratio of 15:1 by weight in fertilizer was chosen to keep the phytoplankton in the phosphorus-limited range, generally considered to be where N:P is >7:1 by weight. Precise nutrient additions were made from weighed bags of commercial fertilizer and measured volumes of phosphoric acid. Detailed measurements of water chemistry and outflow from the lake allowed us to construct much more precise nutrient budgets than had been possible in any previous whole-lake studies (Table 9.1).

To measure the effect of fertilization on the extremely low concentrations of dissolved inorganic carbon, the only form of carbon that could be used directly by photosynthesizing algae (Chapter 4), new methods had to be developed. At the time, dissolved inorganic carbon in lakes was usually calculated, using

TABLE 9.1: *A summary of phosphorus (P) and nitrogen (N) inputs to Lake 227 as fertilizer.*

| Year | P input (g/m²/year) | N input (g/m²/year) | N:P (weight) |
|---|---|---|---|
| 1969 | 0.34 | 5.0 | 14.4 |
| 1970-1974 | 0.48 | 6.2 | 13.0 |
| 1975-1989* | 0.45 | 2.2 | 4.9 |
| 1990-2005† | 0.45-0.47 | 0 | 0 |

*The year 1983 was an exception. Only 0.38 g/m²/year of phosphorus was added, the result of early freezing over of the lake.
†The year 1998 was an exception. Phosphorus addition was 0.60 g/m²/year for that season, the result of a technical miscalculation.

acid titrations to obtain the concentration of bicarbonate, then calculating dissolved inorganic carbon using this value, temperature, and pH in predictive equations. This traditional method proved to be too insensitive for our purposes, as the result of low concentrations of bicarbonate and high concentrations of dissolved organic acids, which interfered with the acid titrations. However, talented scientists at the Freshwater Institute solved the problem. Dr. Domenico Povoledo was able to develop a technique to measure dissolved inorganic carbon directly, using a blood gas analyser connected to a gas chromatograph, which was sensitive to about 2 micromoles per litre dissolved inorganic carbon. Later, Mike Stainton simplified the method to reduce the amount of sample needed and the time required for analysis. He also improved the sensitivity, by equilibrating gases directly in a large syringe. Today such methods are in widespread use, but it was not a trivial problem in 1969.

The answer to the carbon-limitation question was swift in coming. Within weeks of beginning to fertilize the lake with phosphorus and nitrogen, Lake 227 developed a huge algal bloom, of mixed desmids and *Oscillatoria* (now *Planktothrix*) *redekei*. Several other species in the lake also increased greatly in abundance (Schindler et al. 1971, 1973; Figure 9.1).

In the process of the bloom developing, the originally low dissolved inorganic carbon concentrations in the lake (<50 micromoles per litre) were drawn by algal photosynthesis to near the limits of detection (<2 micromoles per litre) in midday. There was a strong diurnal cycle. $CO_2$ increased overnight as the result of invasion from the atmosphere and respiration in the lake, then disappeared within a few hours after sunrise as photosynthesis, stimulated by our additions of phosphorus and nitrogen, began to demand dissolved

FIGURE 9.1: *Left: Lake 227 before fertilization, with apparatus for several smaller experiments visible on the lake's surface. Right: Lake 227 in midsummer after fertilization began. Oligotrophic Lake 305 is shown in both pictures. For colour image see p. 329.*

inorganic carbon (Figure 9.2). Clearly, even such low dissolved inorganic carbon concentrations did not prevent algal blooms. After publication of our first few papers and presentations at various scientific meetings, scientific papers and propaganda supporting the carbon limitation hypothesis quietly disappeared. The theory that carbon caused eutrophication never resurfaced.

## New Puzzles

WHILE THE RESULT of the Lake 227 experiment was clear, it raised some interesting secondary questions. The algal bloom in the lake now contained far more carbon than there had been in the lake as dissolved inorganic carbon before the experiment. Where had the carbon come from? Why had Lake 227 responded so clearly to addition of P and N, when laboratory studies using water from Lake Erie and other lakes had not, unless carbon was added? Studies in small bottles also indicated that algal growth was limited by carbon in Lake 227 (Figure 9.3).

Clearly, the algal blooms in Lake 227 had a carbon source other than the DIC in the lake. Possible sources were few: dissolved organic carbon (DOC), which originated from terrestrial soils and excretion by aquatic plants, could be converted to $CO_2$ by bacteria and ultraviolet light. Dissolved organic carbon in the lake was quite high, largely as the result of runoff through forested soils. However, there was no evidence that dissolved organic carbon concentrations decreased after fertilization; in fact, long-term trends show that they have increased slowly over the years. Another possibility was that rapid diffusion of $CO_2$ from sediments or the hypolimnion into the euphotic zone could have supplied the $CO_2$, but calculations based on the density gradient at the

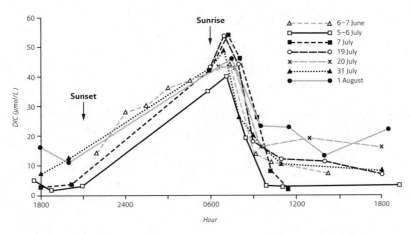

FIGURE 9.2: *The diurnal changes in concentration of dissolved inorganic carbon (DIC) in the epilimnion of Lake 227 at the height of the summer algal bloom, showing the increase overnight caused by respiration and entry of atmospheric carbon dioxide, followed by photosynthetic uptake after sunrise. The daytime concentrations after 10 am are probably overestimates of DIC as the result of contamination of samples by atmospheric $CO_2$ between sampling and analysis, given the technology of the time.*

From Schindler and Fee (1973). Printed with permission from NRC Press.

thermocline and textbook diffusion coefficients for sediment–water exchange indicated that these were unlikely to be significant sources of carbon.

Finally, only the atmosphere could have been supplying $CO_2$ to the lake. Fertilization of the lake with phosphorus and nitrogen stimulated carbon uptake by photosynthesising algae, which reduced $CO_2$ concentrations in the epilimnion to well below atmospheric saturation (Figure 9.4), causing a net flux of $CO_2$ into the lake. Some of the $CO_2$ used in photosynthesis during the day was clearly replenished overnight by respiration and atmospheric input (Figure 9.2). Further research would be needed to determine the roles of respiration and gas exchange as carbon sources.

One of the world's leading gas-exchange experts, Dr. Wally Broecker of Lamont Geological Observatory (now Lamont-Doherty Earth Observatory), was invited to visit the ELA in 1969. His primary interest was in measuring gas exchange in the oceans, where preliminary studies of the invasion of bomb radiocarbon indicated that the oceans were important "sinks" in global carbon budgets, with direct bearing on the prediction of rates of climate warming. Their measurements of gas exchange were restrained by the same methodological limitations as ours. Broecker and his colleagues had made

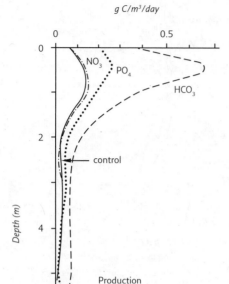

g C/m³/day

Depth (m)

Production
g C/m²/day

| | |
|---|---|
| control | 0.264 |
| PO₄ | 0.530 |
| HCO₃ | 1.139 |
| NO₃ | 0.278 |

FIGURE 9.3: *A test of nutrient limitation of primary production in Lake 227 after fertilization, done in closed bottles like many of the experiments publicized by the Soapers. It is assumed that, if primary production increases following addition of a nutrient, that nutrient is limiting production and that the largest increase indicates the most limiting nutrient. The above result shows the fallacy of the Soapers' argument: despite the fact that carbon was the most limiting nutrient in small bottles, the lake had become eutrophic because of adding phosphorus and nitrogen, as described in the text. The limitation shown by bottle assays was in response to over-fertilization with phosphorus and revealed nothing of value to reducing eutrophication.* From Schindler (1971b). Printed with permission from Blackwell Publishing.

initial measurements using carbon-14 produced in atmospheric bomb testing. With the era of bomb testing over, they were experimenting with methods employing radon, a naturally occurring radioactive gas that could be measured with great accuracy, to estimate fluxes of other gases, including $CO_2$. Radon measurements of gas exchange had a number of advantages. Radon is present naturally in lakes as well as the oceans as the result of a continuous supply from the radioactive decay of uranium-238 in soils and lake sediments (Table 9.2). It could be stripped from a large volume of water and concentrated

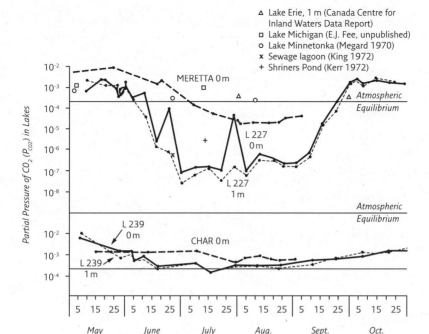

FIGURE 9.4: *The concentration of carbon dioxide in several lakes, including Lake 227. Upper panel: Eutrophic lakes, including several in the United States and one (Meretta) in the high Arctic, showing how addition of phosphorus and nitrogen causes algae to remove carbon dioxide to below atmospheric concentrations. Meretta Lake received the sewage from the North Base at Resolute Bay. Lower panel: The concentration of carbon dioxide in oligotrophic lakes. Lake 239 is at the Experimental Lakes Area, Char Lake is in the high Arctic, the site of an International Biological Program study. It is within a few miles of Meretta Lake. The lakes are often above atmospheric saturation with $CO_2$, the result of decomposition of terrestrial organic matter and algae in the lakes.*

From Schindler et al. (1975). Reprinted with permission from (Productivity of World Ecosystems) © (1975) by the National Academy of Sciences, Courtesy of the National Academies Press, Washington, D.C.

in a small chamber where highly sensitive radiation counters could be used to measure its abundance. Radon is continuously generated at constant rates from radioactive decay in a lake's sediments, diffusing into water. It then escapes to the atmosphere, which had a concentration near zero, due to the rapid radioactive decay of radon to lead-210 (half-life of 3.825 days). The lead returns to Earth with precipitation, forming the basis for other measurements in lakes. The rapid radioactive decay of radon in Lake 227 meant that profiles of radon concentration in the lake measured several days apart could be considered as independent measurements of gas exchange.

TABLE 9.2: *The chain of radioactive decay from uranium-238 to lead-206 showing radium-226, radon-222, and some of the lead and polonium isotopes that are often used to study lakes and oceans.*

| Element | Half-Life |
|---|---|
| Uranium-238 (U-238) | 4,460,000,000 years |
| ↓ | |
| Thorium-234 (Th-234) | 24.1 days |
| ↓ | |
| Protactinium-234 (Pa-234) | 1.17 minutes |
| ↓ | |
| Uranium-234 (U-234) | 247,000 years |
| ↓ | |
| Thorium-230 (Th-230) | 80,000 years |
| ↓ | |
| Radium-226 (Ra-226) | 1,602 years |
| ↓ | |
| Radon-222 (Rn-222) | 3.82 days |
| ↓ | |
| Polonium-218 (Po-218) | 3.05 minutes |
| ↓ | |
| Lead 214 (Pb-214) | 27 minutes |
| ↓ | |
| Bismuth-214 (Bi-214) | 19.7 minutes |
| ↓ | |
| Polonium-214 (Po-214) | 1 microsecond |
| ↓ | |
| Lead-210 (Pb-210) | 22.3 years |
| ↓ | |
| Bismuth-210 (Bi-210) | 5.01 days |
| ↓ | |
| Polonium-210 (Po-210) | 138.4 days |
| ↓ | |
| Lead-206 (Pb-206) | Stable |

Tests of radon versus $CO_2$ exchange in Broecker's laboratory indicated that reasonable estimates of $CO_2$ exchange could be derived from radon data, simply by correcting radon results for the difference in diffusion coefficients between radon and $CO_2$ (Broecker 1974). Broecker assigned one of his new PhD students, Steve Emerson (now Professor of Oceanography at the University of Washington) to the problem. A problem soon developed: it was necessary to strip radon from 20-litre samples of water to perform an accurate analysis. Several samples, in sealed glass bottles, were needed on each date to estimate detailed lake profiles. Carrying several 20-litre bottles over two portages and three lakes without breaking them was a major logistic feat. To overcome the problem, we added a small amount of radium-226, the uranium daughter that decayed directly to radon-222, to Lake 227, to act as a constant generator for radon gas. The loss of radon from the lake could be calculated as the difference between that supplied by decay of radium and the observed radon concentrations in the epilimnion of the lake. This allowed us to use 2-litre samples for measurements, so that one person could do what had previously required several of us. Emerson (1975a, 1975b) described this work.

When fluxes of radon to the atmosphere were converted to $CO_2$, the results confirmed our earlier deduction (Schindler et al. 1972) that the flux of atmospheric $CO_2$ to the lake was the missing dissolved inorganic carbon source, which allowed phytoplankton blooms to develop in response to phosphorus loading. Clearly, earlier studies conducted in small bottles had ignored an important factor in eutrophication. The radon method went on to provide early measurements of gas exchange between the ocean and the atmosphere, facilitating early estimates of global warming.

### Why Earlier Studies Overestimated Carbon Limitation as a Cause of Eutrophication

As shown in figure 9.3, bottle enrichment experiments in Lake 227 after it was fertilized with P and N showed that the lake was carbon limited. These results were comparable with the experiments that were used by the Soapers to promote the case for retaining phosphate detergents, as discussed in Chapter 7. Clearly, the conclusions based on small bottles were nonsense: The lake showed us that the carbon limitation was the *result* of phosphorus and nitrogen addition, not the cause of eutrophication! Also, bottle experiments excluded the exchange of $CO_2$ with the atmosphere, exacerbating carbon limitation by eliminating one of the key biogeochemical interactions in the natural lake. The results of bottle experiments were revealing *symptoms* of eutrophication, rather than *causes*.

Without whole-lake experiments, inappropriate interpretations of bottle experiments could have led us into expensive and futile attempts to manage carbon, or at least long delays in implementing appropriate policy actions. Suppose that another team of scientists was sampling Lake 227, unaware of our fertilization. They would have run typical bottle bioassays and concluded that, to reduce eutrophication, carbon must be controlled. Young scientists should take note: in recent years, the science of limnology appears to be returning to small-scale approaches that can be precisely replicated. However, such approaches are vulnerable to the very sort of problem outlined above. Small-scale experiments may be precisely replicable, but they are often of questionable realism when applied to the whole lake, because they are missing important components of real ecosystems.

Average algal abundance in Lake 227 was proportional to average phosphorus concentration, despite the carbon limitation indicated by bottle bioassays. The algae also contained normal C:P ratios. Algae had to alter respiration, excretion, or turnover of phytoplankton biomass in order for this to happen. Obviously, the biota of lakes are capable of shaping their chemical environment in powerful ways. This so-called stoichiometry has proven to be important in many ecosystem responses, as outlined later.

A second whole-lake experiment was done to further test the $CO_2$ limitation argument. We added phosphorus, nitrogen, and carbon (as sucrose, a very labile organic carbon source that could be easily converted to $CO_2$ by bacteria) to another small lake, Lake 304. The sucrose was taken up by aquatic bacteria and transformed to $CO_2$ within hours of addition. The additional carbon had no effect on the size of algal blooms beyond what would be expected from adding phosphorus and nitrogen. $CO_2$ in excess of algal requirements was lost to the atmosphere, because $CO_2$ concentrations in surface water usually exceeded atmospheric concentrations (Schindler 1975).

### Is Phosphorus Control Alone Sufficient to Reduce Eutrophication?

THE FIGHT OVER CONTROLLING EUTROPHICATION wasn't over with the demise of the carbon theory. A new series of industry arguments emerged. They claimed that the supply of phosphorus in lakes was inexhaustible, as the result of rapid turnover and release from sediments. The Soapers claimed that such rapid recycling would make it virtually impossible to control phosphorus completely enough to limit plankton blooms, because *"...the*

*phosphates in bottom sediments would simply redissolve to maintain essentially*
*a constant concentration of phosphate in lake waters for hundreds of years"* (Derr
1971, cited in Edmondson 1991). They argued that there would be little point
in controlling phosphorus inputs without controlling other nutrients as well.
Soaper propaganda proclaimed that *"Eutrophication is an immensely complex*
*process that cannot be linked to a single villain such as phosphorus. It can be*
*caused by other nutrients—manganese, iron, zinc, carbon, silica, molybdenum."*
Edmondson (1991) discussed this and other irrelevant arguments made by the
Soapers in considerable detail.

We tested the hypothesis that phosphorus control alone was sufficient to
prevent eutrophication in another whole-lake experiment. Lake 226 is shaped
roughly like an hourglass, with two main basins of similar size and shape
separated by a shallow narrows (Figure 9.5). We installed a sturdy plastic-
coated nylon curtain to separate the two basins, sealing the separation of
the two basins by piling rocks on the bottom edge of the curtain. We then
added nitrogen and carbon to both basins of the lake, but phosphorus only to
the northeastern (downstream) basin. The plan was to simulate the effect of
sewage inputs, with and without phosphorus removal.

Within a few weeks, the result was clear. There was no significant
increase in algae or change in species in the southwestern (upstream) basin,
which did not receive phosphorus. In contrast, the northeastern basin, which
received phosphorus in addition to nitrogen and carbon, produced huge algal
blooms (Figure 9.5). To regulators confused by the Soapers' slick propaganda,
the aerial photograph of Lake 226, first published in *Science* (Schindler 1974),
was clear evidence of the need for phosphorus control. The picture confirmed
that reducing phosphorus in detergents and removing it from sewage effluents would be effective in controlling eutrophication. As discussed earlier,
this position was already held by Canadian regulators, but the picture was
used many times in hearings on eutrophication control, which were held on
a state-by-state basis in the United States. Eventually, all of the states that
had effluents entering the Great Lakes passed phosphorus control legislation.
Most European countries also adopted phosphorus-control measures.

### The Role of Cyanobacteria (Blue-green Algae) in Maintaining
### Phosphorus Limitation

LIKE LAKE 227, the Lake 226 experiment revealed important secondary
questions of a theoretical nature. In the first five years of fertilization, algal
blooms in Lake 227 had always been dominated by chlorophytes, diatoms,

FIGURE 9.5: *Lake 226 in midsummer 1973. The far basin (L226N) is being fertilized with phosphorus, nitrogen, and carbon. The near basin (L226S) is receiving only nitrogen and carbon and is in a near-pristine condition. The yellow line at the centre of the lake is a waterproof curtain separating the basins. For colour image see p. 330.* From Schindler (1974).

and non-nitrogen fixing blue-greens (Cyanobacteria) chiefly *Oscillatoria* (*Planktothrix*) spp. In contrast, Lake 226N was dominated by filamentous, nitrogen-fixing blue-greens, chiefly *Anabaena planktonica, Aphanizomenon gracile,* and *Aphanizomenon schindleri.* Was this difference the result of differences in natural algal species in the two lakes or a result of the difference in nutrient supplies? In Lake 227, where we wanted to keep the lake phosphorus limited, we had used a high N:P fertilizer (15:1 by weight) and had added no carbon. In Lake 226, where we had attempted to reproduce the nutrient ratios in sewage in our additions, fertilizer contained an N:P ratio of only 5:1 by weight, and sucrose was added as a carbon source. The N:P ratio added to Lake 226 was well below the ratio found in algae (see Chapter 1), and similar to the ratios found in sewage. In both lakes, algae consumed all of the added phosphate, leaving undetectable concentrations dissolved in water. This is important, because high inputs of phosphorus do not necessarily cause high concentrations of phosphate in water. Instead, phosphate is taken up very rapidly by plankton, leaving little in solution. The nitrogen necessary to transform the phosphorus into algal growth in Lake 226N had to be coming from somewhere besides our fertilizer. Based on the Lake 227 result for carbon, we hypothesized that inputs of gaseous nitrogen from the atmosphere could be the needed nitrogen source. Nitrogen-fixing blue-green algae (Cyanobacteria) might be favoured in Lake 226N, because other algal species could not use gaseous $N_2$. To test this hypothesis in 1975, we decreased the nitrogen added in fertilizer to Lake 227 to 5:1, while maintaining the previous level of phosphorus input. Within weeks, nitrogen-fixing Cyanobacteria began to increase in Lake 227. By midsummer, they had become dominant in phytoplankton biomass. Studies by Bob Flett (Flett et al. 1980) showed that fixation of atmospheric nitrogen was indeed allowing N-fixers to thrive and outcompete species that could not fix atmospheric nitrogen. *In short, carbon and nitrogen limitation were symptoms of overfertilization with phosphorus but not reliable indicators that carbon or nitrogen control were necessary to reduce eutrophication.* As we shall see in Chapter 13, this same confusion in interpreting the relationship between nutrient limitation and controlling nutrient loading frustrates attempts to control estuarine eutrophication in many cases. Unfortunately, although small-scale bottle and mesocosm experiments allow limnologists to assess nutrient limitation, they tell us little of importance with respect to controlling nutrient loading (Schindler 1998).

It was now clear that the atmosphere had an extremely important role to play in the nutrient biogeochemistry of lakes. By "topping up" carbon and nitrogen supplies as necessary, the atmosphere maintained proper carbon:nitrogen:phosphorus ratios for algal growth. All species of algae could enhance intake of carbon from the atmosphere by removing $CO_2$ in water via photosynthesis, thus maintaining surface water concentrations well below atmospheric saturation. Nitrogen, as gaseous $N_2$, also invaded the lake, but only species capable of nitrogen fixation could utilize it. Therefore, when the N:P ratio was low, nitrogen fixers were able to out-compete other species.

Once fixed by algae, much of the entering carbon and nitrogen remained in a lake. As nutrient concentrations increased, they slowly returned lakes to a phosphorus-limited state (Schindler 1977). Regardless of the ratios of phosphorus:nitrogen:carbon that were added to ELA lakes, the nutrient ratios in algae tended to remain constant, probably because they are biologically determined. *Although additional carbon and nitrogen could be drawn from the atmosphere to maintain the necessary proportions, phosphorus could not, because there is no significant gaseous form of phosphorus in the atmosphere.* Under natural conditions, small amounts of phosphorus enter lakes with precipitation, pollen, and other debris from the catchment but only enough to allow the lakes to maintain their natural oligotrophic state.

Clearly, the classical view that lakes are microcosms isolated from their surroundings was erroneous. This was suggested by W.H. Pearsall before World War II, but studies in the 1960s by Richard Vollenweider, Peter Dillon, Gene Likens, Herb Bormann, and others first showed clearly that lakes respond to changes in nutrient yields from their catchments. Similarly, the prevailing view that the biology of lakes is simply a passive reflection of the lakes' chemistry was destroyed by the experiments in lakes 226 and 227. By their metabolism, changes to species composition, and evolution, the biota of ecosystems are capable of modifying the chemical environment as necessary.

Studies in lakes 227 and 226 show that lake–atmosphere interactions are also important modifiers of the biological processes in lakes. The biota of lakes are by no means passive; they are able to draw upon the atmosphere to correct imbalances in carbon and nitrogen, causing important changes in the lakes' chemistry. The intimate connection of lakes with their watersheds and airsheds has greatly changed the way in which limnologists think about protecting lakes from pollutants of all sorts, not just nutrients.

In summary, the ELA results showed several reasons why control of phosphorus alone is sufficient to reduce eutrophication in most lakes. Atmospheric exchange of $CO_2$ and $N_2$, stimulated by biological activity allowed lakes to correct carbon and nitrogen deficiencies in order to maintain phosphorus-proportional growth. Of the three major nutrients, only phosphorus, with its lack of a significant atmospheric pathway, was incapable of correcting nutrient deficiencies in lakes by natural biogeochemical processes.

## Recovery from Eutrophication

MOST OF THE NUTRIENT EXPERIMENTS in the ELA were terminated by the late 1970s. Without exception, phosphorus and chlorophyll *a* concentrations in fertilized lakes declined rapidly after external phosphorus supplies were terminated, as would be predicted from the strong phosphorus limitation exhibited in all experiments, and the short residence time of phosphorus in ELA lakes. The return of phosphorus from lake sediments was negligible in every case, perhaps the result of high demand for the element by microbiota at the sediment surface and the binding provided by high iron content of sediments in the area. As a result of the rapid decline in phosphorus, phytoplankton populations quickly returned to normal. The only long-term effect of the nutrient addition was a much larger than normal population of lake whitefish in Lake 226N, which required several years to readjust to the lower production of the lake after fertilization ceased (Schindler et al. 1993).

However, in many other lakes, recovery has not been so successful. We will treat this topic in more detail in Chapter 12.

## The Effect of Lake Flushing

THE EARLIEST MODELS relating phosphorus loading to eutrophication were based on the mean depth of the lakes. Based on chemostat theory, Vollenweider realized that more accurate models could be constructed if the water residence times of lakes were included. At its simplest, the water residence time of a well-mixed lake is the lake's volume divided by its outflow per unit time. Typical water residence times range from less than 1 year to over 100 years. Lakes that have longer water renewal times retain a higher percentage of incoming nutrients.

Several models incorporating these concepts were developed by Dillon and Rigler (1974), Vollenweider (1976), and Schindler et al. (1978). These models are closely related. DWS's model is the simplest and is used here to illustrate the basic concept of water renewal (Figure 9.6). Unless return of phosphorus from sediments is of overwhelming importance, the retention of phosphorus in a lake, and the standing crop of algae, are simple functions of the input of phosphorus divided by outflow volume. Vollenweider also developed equations that incorporated corrections for phosphorus return from lake sediments. For simplicity, the return of phosphorus from sediments is not included here but is discussed in Chapter 12. In many lakes, such corrections are necessary. In particular, there have been recent improvements for models of shallow lakes, but the principles outlined here are still fundamental.

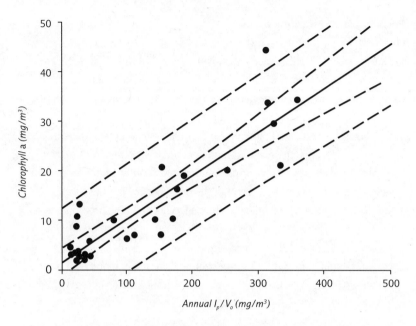

FIGURE 9.6: *The relationship between phytoplankton chlorophyll* a *and phosphorus input divided by outflow volume for ELA lakes.* $I_p$ *is the annual input of phosphorus to a lake.* $V_o$ *is the volume of the lake's annual outflow.* From Schindler et al. (1978). Printed with permission from NRC Research Press.

## The Effect of Trophic Cascading on Eutrophication

SEVERAL AUTHORS DEDUCED that the magnitude of eutrophication was affected by the composition of a lake's biological community as well as nutrient loading and water renewal. In general, larger north-temperate lakes have four trophic levels in pelagic regions. Piscivorous fish feed on smaller fish that feed on filter-feeding zooplankton that consume algae. When piscivores are abundant, smaller zooplanktivorous fish are controlled, so that zooplankton are abundant, and algae are scarce. However, when piscivores are removed, smaller fish flourish, filter-feeding zooplankton become scarce, and algae are numerous (Figure 9.7). This "flip" in ecosystem state from low algal conditions when there is an even number of trophic levels to a high-algal phase with an odd number of trophic levels was termed the "trophic cascade" by Steve Carpenter, Jim Kitchell, and their colleagues, when they reviewed the evidence (Carpenter et al. 1985). Subsequent research has shown that large *Daphnia* must be present in order for a strong "cascading" effect to occur. They are large enough to be of interest to sight-dependent planktivores and

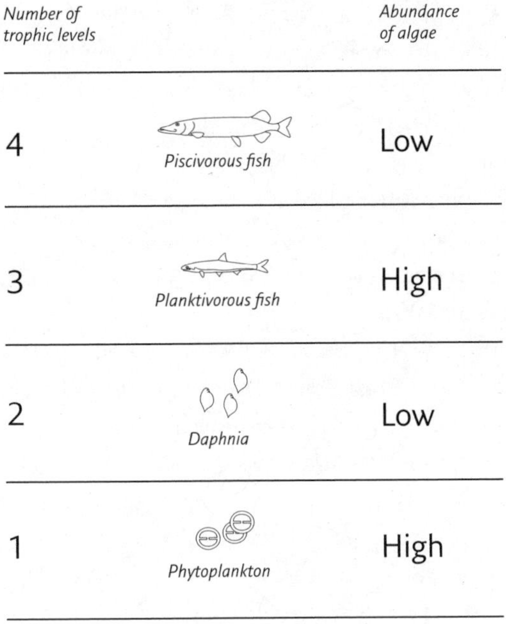

| | | |
|---|---|---|
| 4 | Piscivorous fish | Low |
| 3 | Planktivorous fish | High |
| 2 | Daphnia | Low |
| 1 | Phytoplankton | High |

FIGURE 9.7: *The trophic cascade. At a given nutrient loading, when an even number of trophic levels is present, algal abundance will be in a low phase. If there is an odd number of trophic levels, algal abundance will be in a high phase at the same nutrient loading because of weak grazing pressure by herbivorous zooplankton. Typically, the lakes that humans foul with nutrients also have their piscivore populations overfished, moving the lakes from a low-algal phase to a high-algal phase, aggravating the eutrophication problem.* Diagram by Brian Parker and Lara Minja.

are the only zooplankton capable of multiplying fast enough and grazing effectively enough to reduce algae in most freshwater lakes.

The lakes used for experiments at ELA were too small to contain piscivorous predators, i.e., they represent the high-algal phase in Figure 9.7. All contained populations of several species of minnows, typically the fathead minnow (*Pimephales promelas*) and one or more species of dace, chiefly *Phoxinus* and *Margariscus* species. Lake 226 also contained lake whitefish (*Coregonus clupeaformis*), which feed upon zooplankton and benthic invertebrates. In short, all of the lakes used for nutrient experiments had effectively three trophic levels, i.e., responses would be expected to be at the high-algal phase.

Until the early 1990s, *Daphnia* in Lake 227 were represented only by a few small *D. galeata mendotae*. Small copepods, rotifers, and *Bosmina* dominated the zooplankton. Phytoplankton were dominated by nitrogen-fixing *Aphanizomenon schindleri* (Kling et al. 1994). A thriving minnow community was also present, with several zooplanktivorous species. The estimated minnow density was 105 kilograms per hectare, several times the 10–15 kilograms per hectare found in unfertilized lakes of the area that had no piscivorous fishes. The much higher abundance of minnows in Lake 227 illustrates that phosphorus addition can increase production at higher trophic levels. There were no piscivorous species. It was small wonder that large cladocerans were rare in the lake.

In 1993 and 1994, 200 northern pike (*Esox lucius*) were introduced to Lake 227. They had an average weight of 0.67 kilograms, large enough to prey on minnows. They must have considered Lake 227 to be heaven for pike! By the end of summer, 1994, catch per unit effort of minnows had declined to less than 5% of prestocking values. In 1995, few minnows were caught. None were caught in 1996. By 1996, recaptured pike had increased in size by an average of 0.5 kilograms, and there was evidence that they had reproduced. None of the pike that year contained minnows in their stomachs, further evidence of the effectiveness of pike predation on minnows.

The response of zooplankton to the absence of minnows was slow. Zooplankton biomass remained relatively stable in 1994. In 1995, *D. g. mendotae* began to increase. In 1996, a larger species, *D. pulicaria* appeared for the first time. It multiplied rapidly, displacing the smaller *D. galeata mendotae*. Overall, *Daphnia* biomass increased by 1000-fold (Figure 9.8). By early July, *D. pulicaria* formed 98% of the zooplankton biomass.

*Daphnia* grazing caused the disappearance of the Cyanobacteria blooms that had dominated the lake since the N:P ratio in fertilizer added to the lake was first reduced from 15:1 to 5:1 by weight in 1975. In addition to their effectiveness as filter feeders, *Daphnia* contain much higher concentrations of phosphorus relative to nitrogen than either phytoplankton or other common zooplankton species. As a result, they selectively retain phosphorus from the algae that they feed on. Because *Daphnia* selectively retain phosphorus, the ratio of nitrogen to phosphorus excreted by them is very high, largely in inorganic forms that are available for algal uptake (Figure 9.9). In Lake 227, the resulting concentration of N:P was high enough to allow Cyanobacteria to be outcompeted by faster growing species that could not fix nitrogen (Figure 9.10).

In summary, much has been learned about the control of eutrophication in experiments at ELA. However, funding for eutrophication research at ELA has all but disappeared in recent years. Meanwhile, the eutrophication

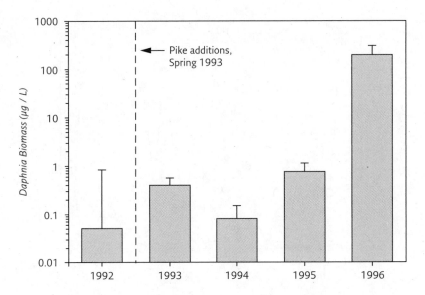

FIGURE 9.8: *Average* Daphnia *biomass in the summers of 1992–1996. Note the logarithmic scale.* From Elser et al. (2000). Printed with permission from Springer Science and Business Media.

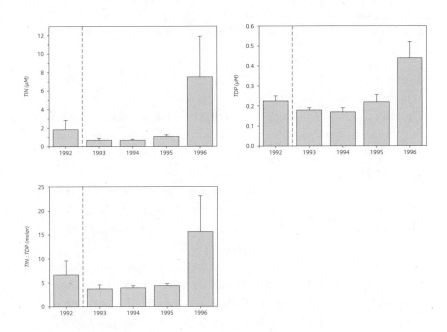

FIGURE 9.9: *The concentration of total inorganic nitrogen (TIN), concentration of total dissolved phosphorus (TDP), and the TIN:TDP ratio of the two in Lake 227. Note the huge increase in the TIN:TDP ratio in 1996, which is the result of selective retention of phosphorus by* Daphnia. From Elser et al. (2000). Printed with permission from Springer Science and Business Media.

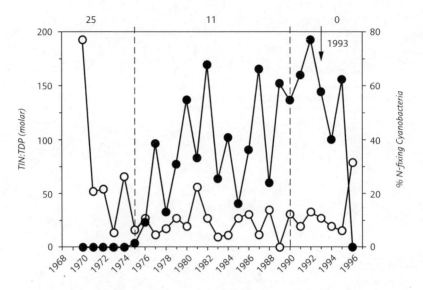

FIGURE 9.10: *The ratio of TIN:TDP and the percentage of algal biomass that consisted of nitrogen-fixing Cyanobacteria. Note the decline in Cyanobacteria in 1996, when TIN:TDP increased due to the high populations of* Daphnia. *The dotted lines are the year that the N:P ratio was reduced from 25:1 to 11:1 (by moles) (1975) and the year the fertilization with N ceased (1990). The arrow represents the year of introduction of northern pike (1993).*
Modified from Elser et al. (2000). Printed with permission from Springer Science and Business Media.

problem in North America has slowly returned as the result of increased human population. Increased intensity of land use, including higher rates of fertilization, increased livestock rearing, trophic cascades set off by depleted predatory fish populations, and reduced flushing of lakes as the climate becomes warmer and drier have also contributed (Schindler 2001). Increasing concentrations of phosphorus, coliform bacteria, nitrogen, and other ominous signs of eutrophication are occurring in freshwaters throughout North America and Europe. Estuaries and coastal marine systems are experiencing similar fates.

In the United States, there were also regulatory problems. The main tool for control of water quality is the Clean Water Act of 1972, with its subsequent amendments. The Clean Water Act gives the U.S. Environmental Protection Agency (EPA) the legal tools to keep lakes "swimmable and fishable," which could easily be interpreted as encompassing nutrient control. However, the EPA, a new agency at the time, chose to concentrate first on toxins and pathogenic organisms, which were arguably even more important 35 years ago. States were allowed to make their own rules, as long as they were not more lenient than the Clean Water Act. Some states took nutrient control seriously, whereas others did not. Although the regulation of toxins and pathogens greatly improved water quality, they left waters in many U.S. states without protection from nutrient enrichment. In the past two years, the EPA has released new criteria for nutrient control to the states, with instructions to adopt meaningful regulations to control nutrients in U.S. waters. A major beneficial change should result. In the next chapter, we shall examine changes to the freshwater eutrophication problem that have resulted from human activities of the past few decades. It is already clear that eutrophication will be one of the foremost environmental problems of the 21st century, a result of the imbalance between the unsustainable growth in human population and industry, and the implementation of regulations to control nutrients. The Algal Bowl has developed much as JRV predicted over 30 years ago, despite the advancements in scientific understanding of the problem. However, the sources of nutrients and other causes of the problem have changed, and the area affected by cultural eutrophication has expanded westward in North America.

# 10 | Changes in the Eutrophication Problem Since the Mid-20th Century

*The state of knowledge of eutrophication in the early years of the 21st century, when eutrophication has evolved to a complex problem resulting from the cumulative effects of nutrient inputs, climate warming, reduced water flows, damage to aquatic communities, and land-use changes in the catchments of lakes*

RICHARD VOLLENWEIDER'S important 1968 review focused attention on phosphorus to solve the freshwater eutrophication problem, as described in Chapter 9. His later deduction that water renewal was as important as nutrient input in formulating models to predict the eutrophication of lakes was an important second step. Curiously, before Vollenweider's contributions, most limnologists had regarded lakes in isolation from their surrounding landscape, perhaps convinced by S.T. Forbes's 1887 paper "The Lake as a Microcosm." However, by the mid-20th century, the evidence was overwhelming that both catchments and the atmosphere had important influences on lakes. Vollenweider's models included both external nutrient sources and water flows, which were driven by atmospheric precipitation and runoff from the catchments of lakes. They formed the basis for eutrophication management in freshwater in the latter half of the 20th century and continue to do so today. As we will discuss below, they have also contributed to understanding of how climate change will affect the eutrophication of lakes.

Whole-lake experiments at Experimental Lakes Area (ELA) (Chapter 9) showed that phosphorus control was effective and practical. They revealed how phosphorus control would significantly reduce eutrophication in most lakes, whereas nitrogen or carbon control would not. As a result, the formulation of detergents in many countries was changed to reduce or eliminate phosphorus. Larger cities on the St. Lawrence Great Lakes and at many other locations also adopted "tertiary treatment" to remove phosphorus from sewage effluents, by precipitation with iron, aluminum, or other cations that formed poorly soluble complexes with phosphate. Together, these measures reduced phosphorus inputs from point sources to the Great Lakes and other important lakes and rivers, significantly reducing the eutrophication problem. Currently, state-of-the-art waste treatment plants can achieve removal efficiencies for phosphorus that are better than 95%. Overall, the control of eutrophication seems like a success story. Certainly, the speed with which scientists were able to identify how to solve a major environmental problem is something that limnologists should be proud of.

But in the 21st century, eutrophication haunts us once again. Cities have grown larger, so that even with high-efficiency phosphorus removal, inputs of nutrients to freshwaters from urban areas are again increasing. Runoff from streets and storm drains during storm events is still not treated in many cities, and most cities are expanding. Paved roads provide near-perfect conduits for transferring pet excrement, lawn fertilizer, and other nutrient sources, as well as herbicides, pesticides, toxic metals, and other contaminants from urban and suburban areas directly to freshwaters. In the case of large urban centres around lakes, nutrient inputs from storm runoff can approach those from sewage. Golf courses are expanding in numbers, using large quantities of fertilizer and other chemicals. Intensive livestock culture; conversion of forested land and wetlands for agriculture, pasture, or urban development; and increasing fertilizer use have replaced

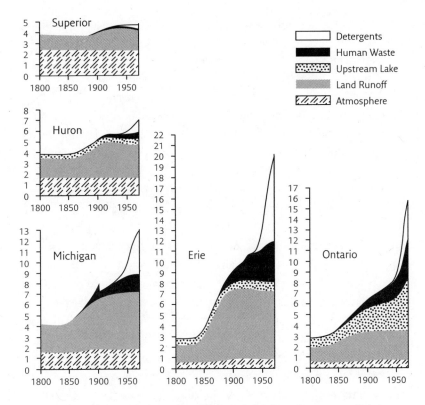

FIGURE 10.1: *Sources of phosphorus to the Great Lakes from 1800 to 1970. Units are thousands of tons of phosphorus per year.* From Chapra (1977). Printed with permission from American Society of Civil Engineers.

or supplemented human sewage as major sources of nutrients. Suburban development and cottage development near lakes and rivers are typically not served by modern sewage treatment. Leaking septic tanks, changes in land cover that allow erosion, pet excrement, and lawn fertilizer are some of the typical sources of nutrients from suburban or cottage developments. These non-point sources are much more difficult to control than detergents and municipal sewage.

Ironically, scientists could see much of this coming at the time when detergents were being eliminated. Land use was clearly identified by PLUARG (for Pollution from Land Use Activities Research Group, of the International Joint Commission) as a major contributor to nutrient inputs in the lower Great Lakes (Figure 10.1). However, bureaucrats and politicians of the day were anxious to declare that the eutrophication problem had been solved, and little money was allocated to fund continued work on the topic. Even the ELA was considered a "sunset," program, the bureaucratic euphemism for closure within a few years.

Fortunately, although government money for scientific work on eutrophication was severely reduced in North America, studies with other objectives continued to add to our knowledge of the problem. Some projects were "bootlegged," carried out by people in their spare time, or continued within other research objectives, such as the Lake 227 experiment. In other cases, "pure" science programs in universities added key knowledge, such as the combined nutrient—trophic cascading experiments done by University of Wisconsin scientists working in northern Wisconsin and Michigan.

The situation in Europe was somewhat better managed. Continued support for eutrophication studies has led to numerous long-term studies of the recovery of lakes from eutrophication (or "oligotrophication" as it is sometimes called). Reducing nutrient inputs, restoration of damaged food chains, and several artificial measures were studied. We shall discuss some key findings later. The October 2005 issue of *Freshwater Biology* and Scheffer et al. (1993) are excellent examples of recent European work.

As a result, limnologists have continued to learn about eutrophication, slowly improving the tools necessary for repairing culturally eutrophied lakes. In the remainder of this chapter, we set out some of the most important of the many issues that have emerged to complicate the eutrophication problem. It is clear that modern eutrophication is the result of the cumulative effects of several sorts of insults to lakes, including climate warming, changes to land use in watersheds, and damage to fisheries.

## The Effects of Drought, Climate Warming, and Land-Use Change on Eutrophication

CLIMATE WARMING plays an important role in modern eutrophication. Many areas of western North America have already warmed by 1–4°C in the 20th century. This has caused ice-free seasons to become longer and summer water temperatures to become higher. As a result, evaporation has increased, silently removing water from lakes, rivers, and aquifers. Of course, the atmosphere must become saturated with water vapour eventually and lose water as precipitation, but this is unlikely to happen exactly where the evaporation occurred. Also, winter snowpacks and glacial water sources are dwindling. As a result, water is not available in the spring and

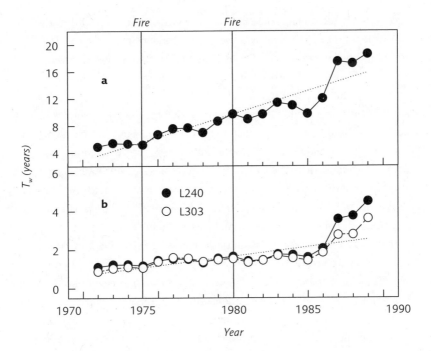

FIGURE 10.2: *Changes in the water renewal times ($T_w$) of three lakes at ELA during a 20-year period when climate warmed 1.6°C and rainfall was below normal. The vertical lines marked "fire" indicate when forest fires occurred in the area. The 1980 fire burned almost all of all three catchments. Water renewal time in panel a is Lake 239.*

From Schindler et al. (1996b). Copyright (1966) by the American Society of Limnology and Oceanography, Inc.

summer, when lake levels are usually replenished and reservoirs filled. The implications of reduced water flows for eutrophication from point sources are obvious, given the importance of water renewal outlined in Chapter 9.

Warmer lake waters also mean that less oxygen is available, because the solubility of oxygen declines as temperature increases. Also, warmer temperatures cause rates of decomposition to increase, consuming oxygen more rapidly. Thus, climate warming can intensify the normal chemical and bacterial processes in lakes. Also, warmer weather usually causes lakes to be free of ice and snow for longer periods, allowing longer periods of stratification and, hence, greater oxygen depletion in the hypolimnions of lakes.

As an example, the water renewal times of lakes in the ELA increased fourfold with a 2°C increase in temperature from 1970 to 1990 (Figure 10.2). The increased renewal times were, in part, the result of lower precipitation and, in part, the result of increased evaporation. Obviously, less water entering and leaving a lake means that any point sources of nutrients entering lakes receive less dilution. The percentage of incoming nutrients flushed from lakes' outflows decreases, and the percentage retained in the lakes increases.

Vollenweider's models generally work well for lakes where eutrophication is caused largely by point sources, where nutrient additions to lakes are largely independent of water inflow. However, where non-point sources are concerned, the situation is different. Here, nutrients reach lakes irregularly, usually during large rainstorms or snowmelts. In general, during dry periods, there is less transport of nutrients and other chemicals to lakes. For example, in natural lakes at the ELA, drought and 2°C of climate warming between 1970 and 1990 caused a slight oligotrophication of Lake 239, as a result of declining inputs and concentrations of phosphorus (Schindler et al. 1996b). The situation is even more complicated where natural vegetation in catchments has been replaced by pastures, croplands, or urban areas. Forest canopies appear to absorb some of the water and energy of falling rain, slowing the transport of water and nutrients, and when forest cover is removed, water is transported more rapidly, carrying more nutrients with it. Also, in lakes with very fast water renewal (days to weeks in some wet regions), nutrients can be washed through lakes without causing fully developed algal blooms or without much loss of nutrient to lake bottoms via sedimentation. Thus, climate warming and land-use changes can greatly complicate the management of lakes where both point and non-point nutrient sources occur. More studies are needed to construct the necessary models to allow reliable eutrophication in all waterbodies, but it is clear that all models must include both phosphorus inputs and water renewal rates.

### How Climate Warming Might Affect Eutrophication

The focus of reports by the media on climate warming is usually on global average predictions of temperature and precipitation. Average global temperatures are heavily influenced by the 70% of the Earth's surface covered by the oceans. The oceans, with their enormous heat capacity, warm much more slowly than land areas. Also, particularly at northern latitudes, dwindling of snow and ice causes dramatic increases in absorbance of solar radiation, because white reflective surfaces change to darker ones that absorb incoming solar radiation. This further accelerates warming. In general, warming of mid-continental land areas is expected to be greatly in excess of global averages. Increases in polar regions are predicted to be still more extreme.

To most Canadians and northern Europeans, the prospect of a warmer climate and more or less normal precipitation sounds wonderful. People forget that, if all other conditions are equal, increased temperature will cause increased evaporation (E) and evapotranspiration (Et), the combined effect of evaporation and water loss by plants through transpiration. Evaporation from water surfaces and evapotranspiration in the catchments of lakes are important parts of the picture for freshwaters. Wind, humidity, and solar radiation also affect evaporation and evapotranspiration, but they are difficult to forecast with existing climate models, so we shall focus on the effects of predicted temperature changes. As an example, we present predictions for the Lakeland Area of northeastern Alberta, where eutrophication is becoming a common problem, as described later for Lac la Biche. Rapid increases in demand for water have accompanied growth of the forest and petrochemical industries. Long-term average precipitation for the area has been about 500 millimetres per year, with long-term evaporation from water surfaces averaging only a few millimetres less.

Human populations in Lakeland have increased with industrial activity. Improved roads have increased ease of access from growing population centres in the Edmonton and Fort McMurray areas, resulting in increasing development of cottages and year-round lakeshore homes. Climate warming has already averaged about 2.6 °C for the area since the mid-20th century. As a result, predicted evaporation from lake surfaces has increased by about 12%, depending on changes in humidity and wind. If the climate warms by this much again, as global climate models predict for the next 30 years, evaporation will be 20–25% higher than in the mid-20th century. By comparison, various climate models predict precipitation changes of from −4% to +13%, with a median prediction of about a 4% increase (Figure 10.3). In short, although models predict slightly more rainfall, the effect will probably be more than nullified by increased evaporation. Lakeland is likely to become a much drier landscape.

The problem is even greater for drier landscapes of the western prairies. Here, precipitation and E + Et are in precarious balance at the best of times. In the past 30 years, a warming trend of 2–4 °C has caused water levels to decline and salinities to increase. Some lakes have become closed basins. For example, Lake Wabamun, one of the most popular lakes near

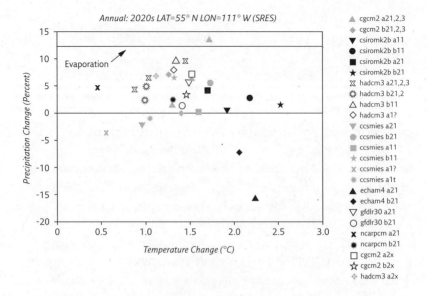

FIGURE 10.3: *Climate change predicted for the Lakeland Area of Alberta from various scenarios of major Global Climate Models. We have added the line marked "evaporation" for the amount of precipitation that would be needed to compensate for the increase in evaporation has occurred from the mid-20th century to the present, calculated from the temperature increases. By the middle of the 21st century, the increase in evaporation will double, based on predicted increases in temperature.*

Redrawn from data on the website http://www.cics.uvic.ca/scenarios.

Edmonton, has not had outflow since 1992, decreasing in volume by 17% and increasing in salinity by about 21% (Schindler et al. 2004).

Another climate problem of the mountainous regions of the Canadian west will be reduced water flows from glaciers and snowpacks. All of the major rivers of the Canadian west originate in the Rocky Mountains, where precipitation is much higher than on the prairies, and all of the glaciers in southern Canada occur. Mountain water is particularly critical to the midsummer river flows throughout the prairies, after snowpacks at lower elevations have melted. Mountain waters also replenish many prairie aquifers, driven by the difference in elevation. Some of the water that is drawn from wells left the mountains hundreds of years ago.

Glaciers of the Rockies are melting rapidly. The termini of the glaciers that recharge the major rivers of the prairies have receded about 1.5 km in the 20th century. Most have lost 25% or more of their mass. Melting glaciers are important components of mid- to late-summer river flow, when annual snowpacks have melted. For example, in a dry summer, 50% of the water in the Bow River at Banff can originate as glacial melt.

One effect that glacial melt has on eutrophication in mountain rivers is via the turbidity that melting ice adds to rivers. This prevents light from reaching the rocky bottoms of rivers where growth of algae occurs below sewage effluents. As a result, unsightly algal growths have occurred primarily in late autumn, when glacial melt dwindles, allowing rivers to clear. However, as glaciers dwindle, turbidity will decline, and clear water will occur earlier in the season. This has already been observed in some years at locations in the Canadian Rockies (Bowman et al. 2007).

Also, as climate has warmed and continues to warm, a higher proportion of winter precipitation in the mountains and foothills falls as rain, particularly at southern latitudes and lower altitudes. Much of the snow that would have accumulated under mid-20th century conditions now seeps away during increasing periodic midwinter melts. As a result, the normal spring freshet from snowmelt is weakened. For example, models from scientists at the University of Lethbridge predict that the spring water yield from the foothills portion of the Oldman River basin will be little greater than half of the 20th century normal by about the middle of the 21st century.

*Removing Protective Features of Catchments*
Watersheds can be regarded as containing a sort of protective "armour" that damps the effects of nutrient inputs from land use and extreme runoff events. However, most landowners are unaware of these protective features and how they can be compromised by bad land-use planning. We have already mentioned the damping effect of forest cover, which is lost upon land clearing, but there are other problems caused by land-use changes.

Draining and filling of wetlands for agriculture and urban development have exacerbated the effects of climate warming and drought on water supplies and eutrophication. Wetlands once acted as "capacitors" in hydrological systems by trapping water from rainstorms and snowmelt, allowing it to seep slowly into lakes, aquifers, and rivers. Many landowners do not realize the critical roles that wetlands play in reducing pollution. They filter out silt, nutrients, and contaminants, as well as delaying the transfer of water to lakes, streams, and groundwaters, acting as significant recharge areas for aquifers. Similarly, the destruction of near-shore riparian vegetation has reduced the capacity of natural vegetation to trap nutrients and silt before they reach freshwaters. Accelerated human use of water as the result of increased population, agriculture, and industrial growth also contributes to water shortages.

*Drought History*

Finally, there is considerable scientific and historical evidence suggesting that the 20th century, when Europeans immigrated to the Canadian prairies, was an abnormally wet period. The only two droughts of significance were the "dirty thirties" and the drought of 1998–2003. By analyzing the width of tree rings and the changes in salinity-sensitive diatoms preserved in lake sediments, scientists have deduced that earlier centuries usually had three or four droughts. Many were a decade or more in length. For example, Sauchyn et al. (2006) have concluded from tree ring analysis that, during the five decades 1830–1880, the Oldman River basin of southern Alberta flows were below the 20th century average in 41 of the 50 years. Using these same "proxy" indicators of climate, the dirty thirties and the recent drought are puny by comparison (Sauchyn and Skinner 2001). Paleolimnological studies of diatoms have similar conclusions (Laird et al. 2003).

Historical evidence of drought is also available from about 1700 onward. Perhaps best known is John Palliser's diary of crossing the southern prairie provinces in the 1850s. It was so dry that he proclaimed in his 1860 journal:

> This large belt of country embraces districts, some of which are valuable for the purposes of the agriculturist, while others will forever be comparatively useless. The least valuable part of the prairie country has an extent of about 80,000 square miles and is that lying along the southern branch of the Saskatchewan River and southward from that to the boundary line....

It is noteworthy that, according to the proxy analysis referenced above, Palliser's visit would have been preceded by three decades of drought. Other, similar descriptions of prolonged droughts can be found in the journals of early explorers and the Hudson's Bay and Northwest companies. The accounts confirm the timing of droughts suggested by tree ring and diatom studies (Sauchyn et al. 2002, 2003).

The net result of climate warming on eutrophication is to reduce inflows and outflows of lakes, causing increased retention of water and nutrients.

In Chapter 9, we showed that this would cause higher nutrient retention and greater algal blooms.

Reduced inflow of water brings a second problem that can accelerate the symptoms of eutrophication. The algae of most pristine northern lakes are dominated by diatoms, which remove silica dissolved in the water to build frustules that are for all practical purposes made of glass (see Figure 1.2). Most of the incoming silica enters lakes from inflowing rivers and streams. There is very little of the element in rain and snow. If inflows of water are reduced, so are silica inputs (Schindler et al. 1996b). When inputs of P and N increase, diatoms grow more rapidly, depleting dissolved silica by early summer. After that, diatoms are replaced by non-siliceous algae. The N:P ratios in nutrient supplies are also low, if sewage or agricultural wastes are the sources. Under these conditions, nitrogen-fixing Cyanobacteria are generally the diatoms' successors, because of their ability to use atmospheric nitrogen (Schelske and Stoermer 1971; Schelske 1999). Thus, as climate warming reduces silica inputs and causes lakes to become ice-free earlier in the year, the onset of Cyanobacteria blooms can begin earlier in the summer season.

### Non-Point Nutrient Sources

At the time of the first edition of *The Algal Bowl*, the effects of livestock were thought to be small, because most livestock were raised in dry, western areas where there was little water to carry the nutrients in wastes to watercourses. Even so, scientists calculated that about 50% of the phosphorus loading to the Great Lakes would continue as a result of non-point sources, largely from livestock and land-use changes (see Figure 10.1). It was also recognized that such sources would be difficult and costly to control. However, as mentioned above, politicians were anxious to claim victory over phosphorus pollution and impatient to reduce the money spent for eutrophication research, so the problem was ignored.

The numbers of livestock in the United States and Canada increased rapidly in the late 20th century, as the result of foreign markets for meat. Much of it destined for developing countries of the Far East. Much of the increase in production has been in confined animal feeding operations (CAFOs).

Land use has changed from forest or grassland to pasture or field. Fertilizer use and livestock culture have intensified. Even partial clearing of forests and converting it to hayland or pasture can double the nutrient

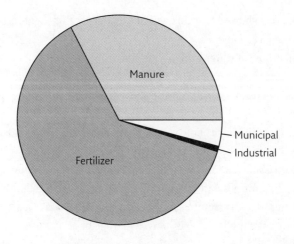

*Total P release: 476,000 t/year*

FIGURE 10.4: *The proportion of phosphorus released to the environment each year by different human activities in Canada at the turn of the 21st century.*
Data from Chambers et al. (2001).

export from a catchment to water. Also, to increase the area available for pastures and fields, riparian areas that protect stream banks and lakeshores have often been destroyed. Over 50% of the wetlands in southern Canada, and even higher percentages in some parts of the United States, have been drained or filled, eliminating "the kidneys" of watersheds where pollutants and silt are normally retained instead of passing them on to aquifers, lakes, and rivers. Pastures and haylands, as well as croplands, are sometimes intensively fertilized to increase production, and the number of animals kept per unit area has increased. In the early 21st century, agriculture in Canada supplies almost all of the nutrients released to the environment by human endeavours, not 50% as it did 35 years ago when municipal and industrial sources were higher but agriculture was less intensive (Figure 10.4).

Millions of animals are kept in CAFOs (Figure 10.5). Some individual CAFOs contain tens of thousands of animals in a small area, often 3000 hectares or less. In some parts of North America, there are many CAFOs within a few miles of each other. The overall numbers of animals in the catchment of a stream or lake can be astronomical. The state of North Carolina has over 12 million hogs, mostly in one small part of the state. The province

FIGURE 10.5: *Some examples of confined animal feeding operations (CAFOs) in Alberta.*

Photographs by Brad Stelfox.

of Alberta had over 6 million cattle in 2004, and there are plans to double the number, although the recent discovery of bovine spongiform enceph-alopathy (BSE or mad cow disease) in Canada and the United States may cause those plans to change. Many states and most Canadian provinces have experienced similar explosive growth in CAFOs. Much of the growth in North America is caused by increasing global demand for meat but also increasing regulation of intensive livestock culture in Europe and Asia, where the problems with air and water pollution were noticed earlier.

Although there are some notable exceptions, the disposal of wastes from CAFOs is generally very primitive. In many cases, manure is simply piled for later application to land or washed with copious amounts of fresh-water into primitive lagoons, where wastes decompose for several months before being applied to land as a liquid spray. Little attempt is made to remove pathogens, nutrients, or other contaminants from the wastes before they are re-applied to the land. In some cases, rates of manure appli-cation far exceed the needs of crops grown on the fertilized land, leaving nutrient residues to accumulate. Odours and airborne ammonia can create enormous problems. Although air and water pollution from CAFOs are now strictly controlled in western Europe, they are still only loosely regu-lated in most parts of North America. We would not dream of simply piling human waste or flushing it into primitive lagoons for a few months, then applying it to the land, yet livestock produce phosphorus and nitrogen in larger quantities than humans, as well as sharing many of the waterborne pathogens that cause human gastrointestinal illness (Tables 10.1 and 10.2). Hormones, antibiotics and other contaminants are also released.

In Alberta, an average beef cow produces 11 times more phosphorus in waste per day than a human. A medium-sized pig produces 10 times more than a human. In other words, a 30,000 head CAFO for cattle produces as much phosphorus as a city of 330,000 people! Based on the numbers of cattle and pigs in 2001, *Alberta received the equivalent (as phosphorus) of wastes from 87 million humans, despite an actual human population of only 3 million.* It makes little sense that we go to great lengths to remove patho-gens and nutrients from human sewage by expensive treatment, but little or no treatment is given to the far greater proportion of wastes from live-stock. It is no coincidence that maps of manure application, phosphorus, and *E. coli* correspond almost exactly.[1]

One problem with large CAFOs is that they import much of the food for livestock from other areas. Nutrients in hay, grain, or other feeds

TABLE 10.1: *Nutrients released by humans and livestock in Alberta in 2001: human nutrient data from Vallentyne (1974) and livestock nutrient data from Chambers et al. (2001).*

| | Population (millions) | Nutrient release (kg/capita/year) | | Tonnes/year | | % total | |
|---|---|---|---|---|---|---|---|
| | | P | N | P | N | P | N |
| Humans | 3 | 0.6 | 4.5 | 1,800 | 13,500 | 3 | 11 |
| Beef cattle | 6 | 6.6 | 11.8 | 39,600 | 70,800 | 76 | 57 |
| Pigs | 1.8 | 6.0 | 22 | 10,800 | 39,600 | 21 | 32 |
| Total | | | | 52,200 | 123,900 | 100 | 100 |

TABLE 10.2: *Relative abundance of enteric pathogens in humans and livestock in Alberta (source of data is unknown).*

| | Human | Cattle | Pigs |
|---|---|---|---|
| Salmonella | 1 | 0–13 | 0–38 |
| E. coli O157:H7 | 1 | 16 | 0.4 |
| Campylobacter jejuni | 1 | 1 | 2 |
| Yersinia enterocolitica | 0.002 | <1 | 18 |
| Giardia lamblia | 1–5 | 10–100 | 1–20 |
| Cryptosporidium parvum | 1 | 1–100 | 0–10 |

are concentrated from several catchments, often in different parts of the country, into the catchment containing the CAFO. For example, DWS recently visited a CAFO in Alberta that fed 36,000 cattle. As the result of several successive years of drought, food for the cattle was not available in western Canada. The operators had to import a hundred railroad cars of corn from the United States per week, to supplement locally grown silage or hay.

The wastes from a CAFO are not usually returned to the catchment where their food is grown, because their high water content makes them uneconomical to transport. As a result, most manure is applied within a 16-kilometre (10-mile) radius of a CAFO. Overall, nutrients are moved from many catchments where food is grown to enrich a few catchments near to CAFOs, often polluting the surface and groundwater in the latter. Manufactured or mined fertilizers are then applied to the catchments where the food was grown to continue growing forage.

Some agriculturists claim that the nutrients added to soil are carefully monitored, and applied only at the rates that are used by successive crops; however, such calculations are often made on the basis of nitrogen, and phosphorus is ignored. As shown in Table 10.1, animal wastes commonly contain N:P at ratios of 2:1 or 3:1 by weight. The crops grown on the manured land have much higher N:P ratios, usually 15:1 to 30:1, so that even if all of the nitrogen applied as fertilizer is taken up by plants and harvested, excess phosphorus is left behind to over-enrich the land. Phosphorus is usually bound to soil particles. It reaches lakes and streams when wind or water erosion moves the soils, often during snowmelt or extreme weather events. Losses of wetlands and riparian areas that once prevented erosion promote the movement of phosphorus to watercourses. In some areas where intensive agriculture has been practiced for several decades (such as southern Wisconsin), excess phosphorus in soils is so high that it is soluble. Even if all sources were cut, the soils would continue to leak phosphorus into surface and groundwater for decades (Carpenter 2005).

In many areas of intensive livestock culture, ammonia released from CAFOs has increased nitrogen in precipitation. Typically, atmospheric ammonia is elevated for long distances around CAFOs, causing increased concentrations in rain and snow. In soil, the nitrification of ammonia to nitrate is followed by leaching of the nitrate into surface water and groundwater. Increased direct applications of synthetic nitrogen fertilizers, particularly anhydrous ammonia, the favourite fertilizer of corn-belt states, cause similar increases in atmospheric ammonia and nitrate in soil. The result is that, in many agricultural areas, nitrate concentrations in surface and groundwater approach or exceed the 10 milligrams per litre of $NO_3$-N that is considered to be the upper threshold for safe drinking water. Some communities have begun to remove nitrate from their drinking water supplies, a costly and complicated procedure. We shall discuss the release of nitrogen, which contributes to marine eutrophication, in Chapter 12.

Like cities, CAFOs have the capability to capture and treat their sewage. There is, however, great reluctance to apply the same costly measures used for humans to animal wastes. Only a few modern CAFOs are taking the effects of their wastes seriously. In one case that we know, hog wastes are composted, rather than flushed into a lagoon. The demand for freshwater is only about one-sixth that of a conventional "slurry" from a lagoon-based CAFO, and the compost contains very little water. The composted wastes are sterilized, dried, and bagged for sale as fertilizer at sites far from the CAFO.

In another case, cattle wastes are composted at high temperatures, generating methane gas that is captured to generate electricity. The resulting nutrient-rich residue contains little water, making it more efficient to transport to areas where the nutrients are needed. However, such efforts to minimize water use and odours and avoid over-fertilization of catchments are still not common. They are not required or even encouraged by the agricultural bureaucracies in most jurisdictions.

CAFOs are only one of several agricultural activities that detrimentally affect water quality. Traditional livestock rearing has also increased in many areas. Often, there are no fences along watercourses, so that animals can wander up and down the banks, causing erosion, destroying riparian vegetation and fish habitat, and defecating or urinating at or near the water's edge. In some cases, wastes from feedlots have been bulldozed onto the ice of streams and rivers, where they are conveniently carried away with spring floods! Pastures are broken and replanted every several years, often using commercial fertilizers. In some operations (as in CAFOs), forage is brought into small areas from miles around, concentrating nutrient residues.

Also of concern is the overuse of manufactured fertilizers. In particular, the Haber–Bosch process made it possible for people to fix atmospheric nitrogen, converting it to ammonia. Fritz Haber received a 1918 Nobel Prize for inventing the process. Thirteen years later, Carl Bosch received the Nobel Prize for upgrading the process to industrial scales, using high-pressure chemistry. This inexpensive process made it feasible for farmers to add much nitrogen inexpensively. In many cases, they only limited nitrogen applications when there was not enough increase in crop yields to pay for the additional fertilizer. Even though the nitrogen uptake might become very inefficient at high application rates, it doesn't cost very much (Figure 10.6), because farmers do not have to bear the costs of treating downstream water for drinking or of reducing eutrophication.

As mentioned briefly earlier, wetlands assist in slowing the flow of water from catchments to lakes and streams. In the process, they filter out

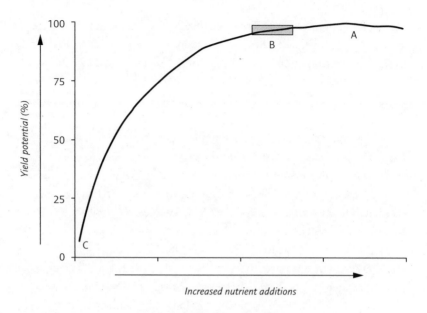

FIGURE 10.6: *A diagram showing the relationship between nutrient addition and crop growth. The low price of nitrogen fertilizer made possible by the Haber–Bosch process allows farmers to profit from adding nitrogen well up onto the asymptotic part of the curve, where the efficiency of uptake of nitrogen by plants is very low (Fixen and West 2002). (A) The asymptote where yield does not increase further with increased addition of nutrients. (B) The range where nutrients are often applied to crops. Little additional yield results, but the low price of nitrogen still makes addition profitable. (C) Initial applications of nutrients result in great increases in yield.* Printed with permission from Ambio, Royal Swedish Academy of Sciences.

eroded soil, retain chemicals, and perform all sorts of other "ecosystem services" that help to protect our waters. Among these services, wetlands are sites where denitrification can remove a substantial part of nitrogen, transferring it to the atmosphere as harmless $N_2$, rather allowing it to flow to rivers, lakes, and groundwaters as nitrate. Wetland ponds are also important breeding sites for waterfowl and amphibians. Treed riparian zones offer shade to cool the waters of streams and ponds for coldwater fishes and other organisms and to protect them from ultraviolet radiation. If these values of wetlands and riparian areas were not ignored, it is likely that they would be protected because of this important "ecological service" that they provide. A few programs have begun to try to educate farmers and ranchers about the importance of maintaining riparian zones and wetlands. Ducks Unlimited and the Cows and Fish program in Alberta are examples. Progress is generally slow, and many watercourses have been so degraded that extensive restoration will likely be necessary. In many cases, it would be more cost effective to pay farmers to protect riparian areas and wetlands, and to use less fertilizer, than to deal with the resulting decline in water quality and ecosystem health.

## Urban and Suburban Development and Eutrophication

Agriculture is not the only culprit in eutrophication. Expansion of urban and suburban areas has occurred widely. In many suburban areas, human sewage is disposed of via septic tanks, which are often poorly installed and maintained. Septic tanks are used for disposing of wastes from about 30% of households in the United States. In cold regions with little snow, septic fields often freeze in winter, allowing sewage to flow overland. They are particularly inefficient for retaining nitrogen, because of the high mobility of nitrate in groundwater.

Suburban development also destroys wetlands and riparian areas. In cities, paved streets act as efficient conduits to transport lawn fertilizer, pet excrement, and many other chemicals to natural waters. Golf courses have proliferated, and many use heavy applications of fertilizer and other chemicals. The resulting "storm drainage" is often allowed to drain to lakes and streams without treatment. In other cases, storm sewers are combined with sewage and directed to treatment plants. Large storms or snow-melt events often increase runoff enough to overwhelm sewage treatment plants, so that the combined effluent is discharged to watercourses without treatment. As the result of the large, sporadic and unpredictable flows caused by weather events, a better course may be to direct storm drainage to ponds or wetlands specifically managed for the purpose of removing nutrients and pathogens before discharge to important waterbodies. Other techniques that can reduce nutrient runoff include installation of roadways without curbs, so that runoff can seep slowly into soils, and use of permeable rather than impermeable surfaces for parking lots and driveways.

There are other pollutants of concern in urban runoff. Herbicides, pesticides, and toxic trace metals are among those that are potentially harmful. For example, wear on automobile tires releases large amounts of zinc. Wear on brake pads releases copper. Automobiles also leach many other chemicals (including toxic metals like cadmium, chromium, nickel, and mercury, lubricants, and coolants), which are carried to watercourses. Intercepting these in soils or wetlands has many potential advantages from preventing water pollution.

*Septic Systems*

Homes and cottages that are too distant from centralized sewage systems usually rely on septic tanks, which simply hold wastes for a few days to allow solids to settle and be bacterially degraded before being pumped or carried by gravity to "septic fields" where porous pipes allow liquids and dissolved nutrients to seep into the soil. A typical septic tank has two chambers, with most of the solids retained in the first chamber, while discharge to the field is from the second (downstream) chamber. Solids build up in the chambers of septic tanks and must be removed periodically to ensure proper functioning. Many septic tanks perform poorly for a variety of reasons. Some soils may saturate with nutrients very quickly, allowing nutrients to seep into groundwater or surface waters. In other cases, tanks are not pumped frequently enough, fields become clogged (or in the north, frozen) allowing effluent to seep over the surface. Some tanks are too small for the volume of effluent they receive. In other cases, maintenance is ignored, and cracked or leaking tanks occur. Septic tanks are not a good solution for heavily developed lakeshores, where effluent flows downhill, generally in the direction of nearby lakes or streams.

It is usually quite difficult to prove quantitatively how much nutrient from septic systems get into waterbodies, but it is clearly some. In the above-mentioned study of Lac la Biche, Mr. Jay White has found that caffeine occurs in areas of the lake near dense summer developments. There are no sources of caffeine except that transmitted through septic systems, showing directly that water from such systems is reaching the lake.

Once municipal authorities find that septic systems are causing severe eutrophication in a popular lake, it is common to have sewage collection and water distribution systems installed around the lake. However, these often cause still further lakeshore development.

There are practical solutions to sewage disposal in such areas. Composting or combustion toilets can reduce organic wastes to small volumes of liquid or solid, which can be easily removed a few times a year. Little or no water is added for waste transport. Of course, the ash or sludge that remains after composting or combustion must be removed from the lake's catchment for disposal.

*Damage to Fisheries, Fish Habitats, and Eutrophication*

In Chapter 9, we discussed experiments in Lake 227 at ELA to test the "trophic cascade." Here, we shall discuss the general role of the trophic cascade in the eutrophication problem.

In the mid-20th century, water quality and fisheries were largely treated as two different disciplines. It was generally recognized that adding nutrients could cause increased fish production, as long as fertilization was moderate enough to prevent oxygen depletion or other adverse secondary consequences of increased algal growth; however, the complexities of aquatic communities were not considered to be a factor in eutrophication. Indeed, in most states and provinces, different government departments still manage fisheries and water quality, as though they were not connected.

A different approach to the problem was taken in eastern Europe and the Orient, where fish were cultured in ponds for human consumption. It is perhaps not surprising that understanding of the factors that allowed nutrients to be transformed efficiently into fish flesh via food chains developed there. The importance of the size and number of grazing zooplankton in controlling phytoplankton, and how the relationship between algae and grazers was affected by fish, were elucidated by Dr. Jaroslav Hrbáček in the Czech Republic. The concepts were translated to North American food chains by Dr. John Brooks and Dr. Stan Dobson a few years later.

One of the first scientists to have experimented with controlling eutrophication by manipulation of fish communities in North America was Dr. Joe Shapiro of the University of Minnesota. He and his students showed that the stirring of sediments by alien species such as carp caused accelerated nutrient return from sediments to overlying water. Later, he proposed that "biomanipulation" of food chains might help to solve the problem of eutrophication where nutrient inputs could not be reduced. However, it remained for Steve Carpenter, Jim Kitchell, and their colleagues at the University of Wisconsin to show in detail the complexity of nutrient–food web interactions and to coin the catchy term "the trophic cascade" for the effects of changing food-web structure on algal abundance.

It may seem bizarre to some that increased fishing pressure and damage to fish habitats are harming our water quality. Most people do not consider fish and water quality to be connected. Indeed, most provinces and many U.S. states have separate agencies for managing fish and water quality. This seems strange in an age when it is clear that fish are an important part of aquatic ecosystems. In Canada, the responsibility for different parts of catchments is a mish-mash of responsibilities. The federal Department of Environment has responsibility for water, whereas the Department of Fisheries and Oceans has responsibility for fish. On the other hand, lands (i.e., catchments) are the jurisdictions of provinces rather than the federal government! Typically, some of the federal responsibilities are delegated to individual provinces, which do not all enforce regulations equally. Often, control of the watersheds is split among departments, for example, departments of forestry, sustainable resources, industry, and agriculture. These artificial separations of parts of aquatic ecosystems and fragmentation of responsibilities confound a true "ecosystem" or "watershed" approach to managing our freshwaters, and this must change.

As discussed in Chapter 9, in the fairly simple communities of northern lakes, there are typically four steps to the pelagic food chain: algae, grazing herbivores, zooplanktivores (chiefly small fish and large invertebrates), and piscivores (pike, *Exos lucius*; walleye, *Sander vitreus*; lake trout, *Salvelinus namaycush*; or smallmouth bass, *Micropterus dolomieu*, are examples). A typical food chain might have a hundred or more species of algae, a half-dozen species of herbivorous crustaceans accompanied by a dozen or more species of small herbivorous rotifers, a handful of minnows and other small zooplanktivorous fishes, and from one to four species of piscivorous predators. Figure 9.7 showed a simple food chain and how the number of trophic levels affects algal abundance. In a simple food chain with strong interactions, an even number of trophic levels causes a low-algal phase at any nutrient loading and water renewal time. An odd number of trophic levels causes high algal abundance at the same nutrient loading and water renewal. Remember that the low-algal phase occurs when the food chain contains an even number of steps. If a step of the food chain is removed, the system flips to a high algal phase at the same nutrient loading.

Of course, there are some caveats. Many lakes contain organism that are "switch hitters," which eat at more than one trophic level. Examples are that some piscivorous fishes will also eat zooplankton. Some calanoid copepods, which are usually thought of as grazing herbivores, can also eat rotifers, smaller crustaceans (including their own juveniles), or protozoans. They can switch from one trophic level to another to optimize their feeding, especially as they grow larger. This weakens their effect on the trophic cascade. As a result, not all lakes show strong trophic cascades. However, in many northern lakes, *Daphnia* dominates the zooplankton, and it is exclusively a herbivore in almost all cases.

The trophic cascade contributes to eutrophication because most sport fishermen like to fish for piscivorous species like walleye, bass, lake trout, and northern pike. If the numbers of piscivorous species are depleted, zooplanktivorous forage fishes increase. *Daphnia*, which are large and slow moving, are ideal prey for these. If *Daphnia* populations are overexploited, the lack of "grazing power" allows algae to flourish. Thus, the "fishing down" of piscivorous species favours a high-algal phase at a given nutrient load.

In populous areas, there is accumulating evidence that we have over-fished piscivorous predator species. One reason is that past fishing regulations were far too liberal. The time required for a predatory fish to reach breeding size was generally underestimated. The "throw the little ones back" mentality persists in many areas today. As a result, the largest fish, which produce the most offspring and are the most efficient preda-tors, are selectively removed, leaving populations with too many immature fish, which are not effective enough as piscivorous predators to keep the "trophic cascade" functioning well. Although catch limits have been slowly reduced over the years, the cuts have usually been political compromises that are insufficient, so that sport fish populations have dwindled...fewer fish, insufficient cuts to limits, so even fewer fish, and so on. The story of the ruination North Atlantic cod (*Gadus morhua*) fisheries is being played out again, in freshwater (Post et al. 2002)!

For example, in cold, heavily fished Alberta lakes, northern pike over the age of 1 year suffer mortalities estimated at about 80% per year. In the first year, it is certainly much higher than that, but good statistics are difficult to obtain. Growth rates are slow, and reproduction doesn't begin until 6 to 8 years of age. The probability of a young pike reaching reproductive age can be as low as 1 in a million! The problem of eroding aquatic communities from the top predators down is so widespread that it is referred to by both marine and freshwater scientists as "fishing down the food web."

The removal of top predators may be difficult to reverse, for a number of reasons. In many Alberta lakes, walleye have been overfished to the point where they have become rare, and their populations have been termed "collapsed" by fisheries scientists. Repeated attempts to restock them have usually failed. As the result of lack of predation, zooplanktivorous fish populations (commonly called minnows, although the juveniles of many sport fishes are also zooplanktivorous) increase greatly. In addition to eating zooplankton, some species of larger minnows can eat fish eggs and even newly-hatched walleye fry, which are little bigger than crustacean plankton. One such species is the spottail shiner (*Notropis hudsonius*) which is commonly called "spawneater" for its habits. Another is the yellow perch (*Perca flavescens*). In Alberta, lakes where piscivorous predators have been overfished, it is common to see huge schools of yellow perch and spot-tails near shore. It is possible that the intensive feeding by these species is enough to keep the recruitment of young walleye low. Newly hatched walleye are not much bigger than an eyelash, ready prey for even small voracious perch.

The explosion in perch and minnows appears to have another undesir-able effect. Numbers of loons, cormorants, pelicans, and other fish-eating birds have increased greatly, because of the abundant minnows. In some cases, the huge increases in bird populations have become a nuisance, killing trees where they roost, defecating on laundry, and cleaning out domestic fish ponds.

At Lac la Biche, where walleye populations have collapsed and large pike are rare, spottail shiners and yellow perch are very abundant. Cormorant (*Phalacrocorax auritus*) populations have increased severalfold, reaching nuisance numbers. They have saturated all suitable nesting sites on islands in Lac la Biche. They have begun to colonize islands on nearby smaller lakes, even if the lakes do not contain fish populations. The cormorants have become "suburban commuters," flying to Lac la Biche to catch fish, returning to disgorge fish for their offspring and defecate in the smaller lakes. As a result, the smaller lakes with expanding cormorant popula-tions are also becoming eutrophic. Residents and fisheries officers are now shooting cormorants and oiling cormorant eggs to try to reduce popula-tions! In short, overfishing has apparently caused the lake to switch to a new steady state, which may be difficult to reverse. Unfortunately, the new steady state where cormorants appear to have replaced piscivorous fish as the top predator in the food chain is one with rare herbivorous grazers, keeping the lake in a high-algal phase. Such regime shifts can be very diffi-cult to reverse (Carpenter 2003). No one seems happy with cormorants replacing walleye and northern pike as the top predator in the lake, but this is not an atypical example of how humans have managed freshwaters.

As a result of the high zooplanktivory by the hordes of minnows and small perch in Lac la Biche, the "grazing power" of herbivorous zooplankton (Chapter 9) is very low. Thus, it is possible that a trophic cascade is exacerbating the effect of high nutrients by keeping the lake in a high-algal phase. The history of eutrophication in Lac la Biche in relation to changes in nutrients and fisheries is discussed in more detail in the next chapter.

It has been hard for sport fishermen, who generally pride themselves for their conservation efforts, to accept the fact that they may be taking too many fish. Some point out that the number of fishing licences being sold is declining, implying that depletion of piscivores is becoming less of a problem. However, most fishermen do not realize that the many advances in fishing technology have greatly increased the efficiency of fishing, more than compensating for any decline in numbers of fishermen. In the mid-20th century, rowboats and small 3–5 horsepower outboards were typically used to access fishing spots. Travel time to fishing spots on larger lakes often took hours. Many lakes in Canada could not be reached at all, unless one was willing to carry a canoe, or to use skis or snowshoes in winter, carrying a backpack containing a hand-auger and a few tip-ups. Today most lakes and rivers in the United States and southern Canada are easily accessible to all because of better roads, cleared lands, networks of logging roads, oil and mineral exploration lines, and improvement in four-wheel drive vehicles and snow machines. Powerful SUVs and all-terrain vehicles can pull large power boats to waters that were inaccessible only a few years ago. Once at the lake, large outboard motors, often over 100 hp, transport fishermen to distant fishing spots in minutes. Good fishing spots can be located precisely using the global positioning system (GPS). Vertical and side-scanning sonar or underwater cameras allow people to detect the presence, number, depth, and size of fish and to assess whether or not they are interested in the lure used, eliminating much of the guesswork that used to be a part of fishing. Modern fishing lures and scents have improved to the point where artificial lures can equal or exceed the performance of live bait. Fishing lines are more invisible and easier to cast, as well as stronger than they used to be. Numerous fishing magazines carry articles by full-time professional tournament fishermen, who tell anglers about recent improvements in ways to efficiently exploit fish populations.

In winter, modern snowmobiles transport fishermen to remote lakes that were once accessible only with difficulty. Instead of chopping a few holes and sitting in one position all day, as was the mid-20th century method of fishing, a "run and gun" technique has been proposed by professional sport fishermen. Power augers allow hundreds of holes to be drilled in the ice, and the same GPS, sonar, and underwater camera used in summer provide information on the location of sites and the presence of fish. If no fish are seen in a half an hour or so, equipment is loaded onto sleds and whisked rapidly to another fishing location, where the procedure is repeated. In summary, improved access and better technology have made fishing much more efficient than it used to be. As a result of the increased fishing pressure, many freshwater fisheries are collapsing in southern Canada. Many in the United States have been long gone.

Some believe that the overexploitation of piscivorous fishes can be solved by making anglers release fish they catch. In some areas, "catch and release" has recently become part of the sport fishing culture. However, the jury is still out on how effective catch and release is at preventing declines in sportfish species. There are numerous problems. Fish that are "played" to exhaustion build up high concentrations of lactic acid, and some studies show that this increases mortality. Bad fish handling practices, keeping fish too long out of water, high water temperatures, and deeply embedded hooks also cause mortality to increase. In some areas, the same fish may be caught several times a year. Mortality rates in catch and release fisheries are sometimes considerably higher than in unexploited stocks, but overall, catch and release seems to have allowed healthy fish populations to return to at least some overexploited waters.

In addition to overharvesting of predatory fishes, fish habitats have been reduced and compromised. One common example is that cottagers on a lake will pull out snags and weedbeds in front of their cottages, removing key habitats for predators and the forage species that they use as food. Typically, fish populations correlate well with the abundance of logs and other "coarse woody debris." At the same time, coarse woody debris is inversely correlated to cottage density (Figure 10.7).

Often, a simple dam is constructed on a lake's outlet, keeping water levels high so that cottagers' boats can be conveniently docked. No thought is given to whether such barriers might affect fish migration or prevent access to spawning or rearing habitats. The effects of reducing water renewal of the lake are also ignored.

Cottagers also typically remove shoreline trees, especially along sight lines from their cabins to the lake. Therefore, shading of shoreline areas,

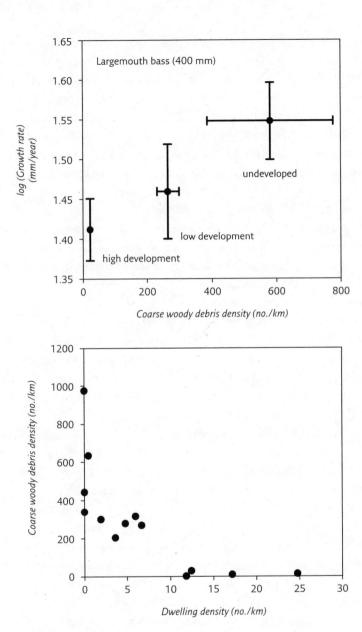

FIGURE 10.7: *Top: The relationship between the density of largemouth bass, a piscivorous predator, and the abundance of coarse woody debris (logs, snags, etc.) in shallow waters of Washington state. Bottom: the relationship between coarse woody debris and the density of lake cottages along lake shorelines.* Examples from D.E. Schindler et al. (2000). Printed with permission from Springer Science and Business Media.

and sources of the dead branches and logs that supply cover for fish, are decreased. Lawns are installed, often with use of fertilizers and herbicides. Many cottages still use septic systems, which are often badly installed or maintained. Pets, sidewalks, and roads provide other sources of nutrients, causing problems similar to those described with suburban development. There are huge developments proposed for several lakes in Alberta, some of several hundred cottages each. Municipal councils will decide the fate of these proposals. These councils usually look primarily at the increases in business expected from such developments, giving only cursory attention to impending problems with water quality or fisheries.

In summary, increased inputs of nutrients, reductions in water flows, and overfishing of predators have caused a "triple whammy" that is eutrophying our freshwaters. The problem is far more complicated than it was in the mid-20th century, requiring management of multiple causes rather than just point nutrient sources (Figure 10.8).

There is a good news part of the story. Humans have most of the scientific basis needed to protect the watersheds of lakes and rivers to prevent eutrophication; however, instead of strategically planning development, we allow it to occur in traditional unplanned ways that inevitably cause lakes to become green (Figure 10.9).

### Eutrophication and the Quality of Drinking Water
Many of the abuses of watersheds that cause eutrophication by allowing nutrients to reach freshwaters also allow contamination of water with pathogens or toxins. The results in the Third World are tragic, with millions of people every year dying or becoming severely ill from diarrhea and disease caused by waterborne agents. However, even in technologically advanced North America, waterborne illnesses are serious. The American Society of Microbiologists estimates that there are 900,000 cases of waterborne gastrointestinal illness in humans in the United States each year and approximately 900 deaths. Some outbreaks have been very severe. For example, in 1993, *Cryptosporidium*, a protozoan that is resistant to chlorination, invaded the water supply of the city of Milwaukee, Wisconsin. By the time the problem was discovered and corrected, up to 400,000 people were infected. Over 4000 were hospitalized, and 50–70 people died. Another example is that of tiny Walkerton, Ontario, where 7 people died and 2300 became ill as a result of contamination of drinking water with virulent *E. coli* O157:H7 and *Campylobacter jejuni*. Another 27 people developed

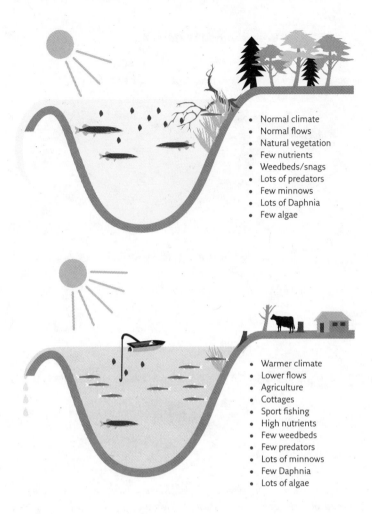

- Normal climate
- Normal flows
- Natural vegetation
- Few nutrients
- Weedbeds/snags
- Lots of predators
- Few minnows
- Lots of Daphnia
- Few algae

- Warmer climate
- Lower flows
- Agriculture
- Cottages
- Sport fishing
- High nutrients
- Few weedbeds
- Few predators
- Lots of minnows
- Few Daphnia
- Lots of algae

FIGURE 10.8: *A depiction of the cumulative effects of nutrient loading, land-use change, climate, and overfishing on the eutrophication of lakes. Top: A natural lake, with abundant top predators, normal water residence time, a catchment protected by natural vegetation, and normal climate. Note the clear blue water, the result of abundant zooplankton (represented here by* Daphnia, *the small red symbols). Bottom: The same lake after human habitation, land-use change, and agriculture have increased nutrient loads; overfishing has removed predators; fish habitat has been destroyed; and climate warming has reduced the outflow, causing increased retention of nutrients. Note that the water has become green, as a result of the combined effects, as described above. For colour image see p. 331.*

Drawing by Brian Parker and Lara Minja.

## Well Planned Watershed

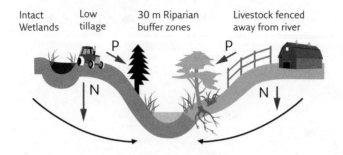

Intact
Wetlands

Low
tillage

30 m Riparian
buffer zones

Livestock fenced
away from river

P

N

P

N

## Poorly Planned Watershed

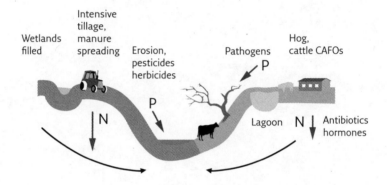

Wetlands
filled

Intensive
tillage,
manure
spreading

Erosion,
pesticides
herbicides

Pathogens

P

Hog,
cattle CAFOs

N

P

Lagoon

N

Antibiotics
hormones

FIGURE 10.9: *Top: A well-planned watershed, where wetlands and riparian zones are protected, and various measures are used to keep nutrients from reaching lakes and streams. Bottom: The traditional way of treating a watershed, allowing development and agriculture to develop without any thought to water protection, is still generally used, because the science to support change is ignored by those who make land-use decisions. For colour image see p. 332.* Drawing by Brian Parker and Lara Minja.

hemolytic uremic syndrome, a serious kidney ailment that can become fatal. The contamination from manure washed into one of the town's drinking water wells during an extreme rainfall event. The Walkerton case provides several useful lessons. Firstly, the source of the bacteria was a farm where manure was applied exactly to government specification. The contamination occurred during a period of high runoff following successive rains. This indicates that current standards for application of manure may not be stringent enough to protect water sources during severe weather events. Indeed, the catchment that supplies Walkerton's water is subject to intensive human use of several sorts. Secondly, the tragedy could have been avoided if chlorination equipment in the contaminated well had been working properly, which it was not. Thirdly, the operators of the Walkerton water treatment plant were deficient, reportedly falsifying data and ignoring indications that chlorine concentrations were insufficient to protect human health. Fourthly, the pipes that distributed the treated drinking water were in poor condition, showing the fallacy of trying to cut corners to save a few dollars. Thus, several "barriers" to contaminants that could have protected humans were violated: watershed protection; treatment plants; trained, competent operators; and a sound distribution system. The message is clear: as long as machines treat our water, some will break down, and as long as humans do the monitoring and engineering, some will make errors. It is obvious that ensuring safe drinking water must include keeping intact the catchments that supply our water, as well as failsafe mechanisms to minimize equipment malfunctions and operator errors.

The true costs of waterborne disease are not considered in our economic accounting of livestock production or suburban development. Instead, these costs are absorbed by the population at large, in the form of increasing water treatment costs or medical expenses that result from gastrointestinal disease. In 2002, the city of Edmonton, Alberta, discovered that *Cryptosporidium* was becoming an increasing problem in its drinking water intake. The source was found to be largely agriculture. Roughly 250,000 cattle live in the catchment of the North Saskatchewan River upstream of the city's water intakes. To prevent an outbreak of gastrointestinal disease similar to Milwaukee's, Edmonton added ultraviolet treatment to its water intakes in 2002 and 2005. The capital costs of adding the treatment were $10 million at one of Edmonton's two water treatment plants.

These costs are low compared with the expense of water treatment for the city of New York. The city was ordered by the U.S. Environmental Protection Agency (EPA) to improve its water treatment by adding a filtration plant. The cost was estimated at $6–8 billion. However, further studies indicated that the same objectives could be met by increasing protection of the 2000 square mile watershed in the Catskill Mountains that supplies drinking water for the city. Estimated costs were $1.5 billion or less than 25% of taking an "engineering approach" (Daily and Ellison 2002). The success of New York's massive experiment is being followed with great interest.

A precautionary approach to prevent problems with water supplies in the first place would undoubtedly have been even less expensive for New York, had anyone recognized the value in watershed protection early in the city's development. An example of such foresight is provided by Portland, Oregon, which receives its drinking water from the 102 square mile Bull Run Watershed on Mount Hood in the Cascade Range. Human activity in the watershed has been very restricted since 1895. There was some timber harvest in the watershed in the 20th century, resulting in 24% of the watershed being logged over time. However, logging was severely curtailed after the mid-1980s and ceased entirely in 1993. Further protection for the watershed was provided by U.S. congressional legislation in 1996. As the result of the extreme protection of its water supply, the City of Portland does not filter its water supply, and the chemistry of drinking water is almost identical to that of precipitation. Annual reports to Portland citizens reveal very low incidence of pathogens as well, and there have been no reports of waterborne illness.

Some communities and individuals draw their drinking water from lakes, ponds, shallow wells, or dugouts. Increasingly, eutrophication of these waterbodies is causing the proliferation of toxin-producing species of blue-green algae (Cyanobacteria). Both hepato- and neuro-toxins have been discovered, sometimes in concentrations that are potentially toxic to humans or livestock.

At present, the most common algal toxin associated with eutrophication of freshwater is microcystin, a potent hepatotoxin produced by *Microcystis* and a few other blue-green genera. In dugouts or in near-shore waters where surface scums of blue-greens can accumulate, it can often reach concentrations of concern. For example, the community of Slave Lake, Alberta, was advised by Alberta Health not to drink the water for several days in 2002, as the result of high microcystin concentrations. Although reliable detection methods have only recently become available for widespread use, there are troubling indicators that the incidence of toxins is increasing. Recent studies on Lake Winnipeg have shown some of the largest blue-green blooms ever seen in North America, often several thousand square kilometres in extent. Associated with these blooms are microcystin and other microtoxin concentrations that are many times the World Health Organization's recommended maximum for drinking water (Figure 10.10).

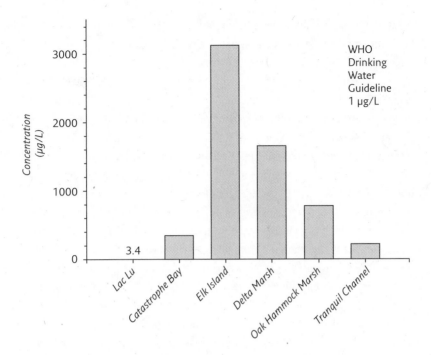

FIGURE 10.10: *Microcystin, a potent liver toxin, in water from several locations in Lake Winnipeg.* Data from Hedy Kling.

Protecting fisheries has several important components. We must protect the predatory species that naturally keep surface waters in the low-algal phase of the trophic cascade. To do this, we must not only accurately regulate the fisheries, but also maintain the habitats where fishes spawn, grow to adulthood, and ambush their prey. We must also prevent as best we can the introduction of alien fish species or of other freshwater organisms that have the potential to disrupt the normal operation of food chains.

Finally, we must consider the limits to human population, to agriculture, and to industry that individual catchments can tolerate. Clearly, as humans and their enterprises increase, the regulation of water use and the treatment of water for human consumption become increasingly complex and expensive. "Full cost accounting" is long overdue, where the costs of water treatment, loss of fisheries, and medical costs of increasing gastrointestinal disease are considered by decision makers rather than just the profits to industries, agriculturalists, fishermen, and developers.

Bureaucracy remains a significant barrier to treating watercourses and catchments in such a comprehensive pattern. Municipal, provincial or state, and national departments are each responsible for certain problems. Even at one level of government, several agencies are involved. For example, a basin plan for a given lake or river in central Alberta would require input from municipal and provincial governments. The federal Department of Fisheries would also have jurisdiction, if fish habitats have been altered, and the federal Department of Environment, if there are industries that discharge pollutants. Even at the provincial level, departments of environment, fisheries, agriculture, sustainable resources, and sometimes energy must be represented. To expedite efficient basin planning, this archaic and cumbersome system must be altered.

In summary, the science of managing eutrophication is now quite well known. However, applying what we know to protect lakes is a complex social and political problem, which must involve careful consideration of processes both within freshwaters and in the catchments that they drain. The networks of streams, lakes, and groundwaters that drain our catchments are the natural "sewers" for the landscape. Yet these sewers are, by definition, where we must draw our water supplies. We must be careful to maintain the "plumbing" of our ecosystems in pristine condition if we wish to continue to drink without trepidation. Figures 10.8 and 10.9 illustrate the features of the necessary comprehensive management and how the continued management of various components of watersheds in isolation has degraded our freshwaters. It is essential that the scientific information that is necessary to make sound water management decisions be communicated to decision makers.

# 11 | The Muddy Archives— Using the Fossil Record to Interpret the History of Eutrophication

*The use of paleolimnology to measure the trajectory of eutrophication and other insults to lakes in the years before modern limnological methods developed*

ONE OF THE DILEMMAS that is usually faced by scientists who wish to assess how much a lake has been degraded by human activity is determining how much a lake has changed from its original state. It is fruitless to try to restore a lake to a condition more oligotrophic than it was under natural conditions. As we discussed earlier in the book, lakes can range from oligotrophic to highly eutrophic for natural reasons, for example, depth, the nutrient content of natural soils, and water renewal. Because most of the methods used by modern limnologists and ecologists were only developed in the mid-20th century and because many lakes were already degraded by that time, limnologists frequently have to resort to indirect means to deduce a lake's original condition.

Fortunately, lake muds or sediments provide a "library" of a lake's history that a skilled paleolimnologist can read. Lake muds are a combination of geological materials (usually fine sands, silts, or clays that are derived from the lake's catchment, or blown in by the wind) and both terrestrial and aquatic biological materials, including many recognizable fossils. Typically, from 1 to 3 millimetres of sediment is laid down each year at the deepest point in a lake, although this can vary considerably. In some lakes, where more or less permanent stratification (called meromixis) allows total anoxia to prevail year round and where deepwater sediments are never stirred by wind or organisms, the sediments accumulate without being mixed. In ideal situations, they can form distinct, recognizable annual bands, known as *varves*. Varved sediments are a sort of "gold standard" for paleolimnologists, because the age of sediments can be determined simply by counting the annual bands, much like tree rings. In ideal cases, even the season when organisms fell to the sediment can be perceived (Figure 11.1). More typically, both wind and burrowing invertebrates stir the sediment surface a little, so that sediments in any layer can be a composite of a few years to a decade or more of biological deposition.

If lakes do not have visible varves, the sediments from different depths in a mud core can still be aged in several ways. In some cases, X-rays of cores can reveal layers not visible to the naked eye, allowing deposition rates to be assessed. Most frequently, radioactive dating methods are used. In Chapter 9, we showed that the radioactive decay of radium produced radon gas that was used to determine the exchange of gases between a lake and the atmosphere. Uranium and its daughter radium-226 are present in most natural rock. As radium decays to gaseous radon-222, the radon escapes to the atmosphere, where it radiodecays with a half-life of 3.8 days. The radon decays to lead-210 (see Table 9.2), which falls back to Earth. The lead-210 that falls into a lake's catchment is generally bound tightly to soils and does not reach the lake except attached to eroding material. That which falls directly on a lake's surface is quickly sedimented to the lake's bottom, with an efficiency of almost 100%. Lead-210 has a half-life of 22.3 years. In theory, the amount of lead-210 in a slice of mud that is 22.3 years old will be half as radioactive as one from the surface. Mud that is 44.6 years old will be a quarter as radioactive as surface sediment, and so on. With modern methods of measuring radiation, it is usually possible to date

FIGURE 11.1: *A freeze-core of lake sediments, showing the fine layering of sediments that characterizes annual varves.* Photograph from John Smol.

sediments accurately back to five to six half-lives, or 120 years or more, until radiodecay leaves too small an amount of radioactivity to be measured precisely. There is one complication: lead-210 is also continuously produced within a lake by uranium series elements in the sediments. This *supported lead* must be subtracted from the total radioactivity to obtain the *unsupported lead* that the lake receives from the atmosphere, as described above. This is easily done, by assuming that the concentration well down in the core (over five or six half-lives) represents supported lead only.

Another method of dating is to look for the peak of cesium-137 deposition, which accompanied the atmospheric bomb tests of the mid-20th century. Knowing that the maximum fallout of cesium was in 1963 allows us to get an estimate of sedimentation rate. There are other small cesium peaks, the most recent when the nuclear power plant at Chernobyl had a large accidental release. Often, these cesium dates are used as a cross-check on lead-210 dates.

For older sediments, carbon-14 dating is often used. The half-life of carbon-14 is about 5000 years. Generally, it can be used to date the entire core from a lake that was formed after the last glaciation, like most of the lakes in Canada, the northern United States, and northern Europe. Of course, there are both geological and biological sources of carbon. The geological sources can be much older, even if they are found in sediments that are only a few years old. Examples are limestone, sandstone, and coal deposits, all of which were formed millions of years ago. Their carbon-14 has long since decayed away. "Dead" (non-radioactive) carbon from such sources can be washed into a lake from its catchment. To get accurate dating, the "dead" carbon must be excluded. Ideally, fossils produced in a lake or its watershed are isolated from the layers of lake mud and dated— invertebrate remains, needles or leaves, and seeds or cones from trees are ideal. In lakes of the Precambrian Shield where limestone and coal are rare, typically the organic matter from a known weight of core can be extracted and assayed directly, because there is no "dead" carbon to interfere.

Biological and geological markers can also be used to age sediments. When European settlers arrived in an area, plants such as European ragweed (*Ambrosia*) or other foreign weeds, accompanied them. It preserves well in sediments, giving a distinct marker that is usually 100–400 years old, depending on when an area was first tilled.

*Tephrae* are layers of volcanic ash that were deposited in lakes following eruption of volcanoes. They can be another useful marker. For example, in western Canada, there were volcanic eruptions of known age during the Holocene. The Mazama Tephra was the result of the eruption of Mount Mazama in Oregon 6800 years before the present (B.P.). Over a cubic kilo- metre of ash was thrown high into the atmosphere, covering thousands of square kilometres. Tephrae are often easy to spot in lake cores because of their distinctive texture. They contain shards of glass, which include distinctive minerals from the region of the eruption. Even in central Alberta, 1600 km away, the Mazama Tephra usually appears as a band in sediment cores about 5–7 centimetres thick. A second, usually thinner tephra is the Bridge River Tephra, dating from 2350 B.P. It is the result of the eruption of Mount Meager, about 180 kilometres north of Vancouver, BC.

Once sediments have been dated, the next step is to determine the biological community that occurred at the time that the mud was depos- ited. Unfortunately, not all organisms in a lake leave reliable fossil remains. One of the classical groups to work with is the diatoms (see Figure 1.2). The pure silicon frustules of many diatoms are usually well preserved for thousands of years and are distinctive for different species. The conditions that favour particular diatom assemblages are well known or deducible for a given region, making it possible to infer what water quality was like over hundreds or thousands of years.

Chrysophyceans also leave silicious fossils in the form of distinctive scales and cysts. The scales are smaller than diatom frustules, and species identification sometimes must be confirmed under a scanning electron microscope. At the time of writing, the cysts are just beginning to be well studied. As a result, chrysophyceans are less frequently used than diatoms for determining the past conditions of lakes.

Other algal groups do not always leave behind recognizable hard fossils. In the past few decades, high-pressure liquid chromatography (HPLC) has made it possible to identify and quantify algal groups by their pigments preserved in lake sediments. These are often used to estimate the abundance of entire algal assemblages in lake muds of known age.

Some invertebrate taxa are also used as indicators. The post-abdominal claws of *Daphnia* are preserved, and often identifiable to species, or at least groups of species. Carapaces of some chydorid Cladocera and of *Bosmina* can also be identified. Unfortunately, the Copepoda and Rotifera, prominent groups in the zooplankton, seldom leave recognizable fossils. Among aquatic insects, several groups can be used. Frequently, the head capsules of chironomid larvae can be identified, and used as indicators of eutrophication. The mandibles of *Chaoborus*, a predatory midge that is common in lakes, are also useful, because the presence of larger species can indicate that a lake is fishless. If fish were present, they would quickly extirpate larger midge larvae. Unfortunately, fish usually leave no fossils that are abundant enough to use in deducing past fish communities.

Of course, the interpretation of fossil records is not perfect. It is rather like trying to deduce what is happening in a hotel room by looking through the keyhole. One might see a pile of clothing on the floor and hear a bed squeaking, but what it means cannot be deduced with 100% certainty! In the past few decades, teams of paleolimnologists with different specialties have often analyzed the same mud cores. If different fossil groups tell the same story, the confidence in interpretations increases.

Eutrophication was one of the first problems studied by early paleo-limnologists. Some of the findings show that eutrophication began very early at some sites. Hutchinson (1969) and Hutchinson et al. (1970) described the sudden eutrophication of Lago di Monterosi, a small volcanic crater lake in Italy, in about 171 B.C. The event that precipitated the event appears to have been the building of a road, the Via Cassia, which increased human visitation to the lake! Recently, Marianne Douglas and John Smol and their colleagues (2004) showed that eutrophication occurred several hundred years ago in the Arctic, when Thule culture humans butchered whales on the shores of a small lake near the seacoast.

## Lac la Biche Alberta, An Example

IN ORDER TO ILLUSTRATE how paleoecology can be used to deduce the original state of a lake and how it has changed, we will use an example from Lac la Biche, Alberta, where DWS and several colleagues have conducted a recent paleoecological study. The lake has a long (by North American standards) history of human occupation, which made tracking the associated changes in the lake of some interest to us.

Lac la Biche (Lake of the Red Deer; a translation of Wa Waskisew Sagihaygan, the Cree name Elk Lake; Figure 11.2) is a large (234 square kilometres), shallow (mean depth 8.4 metres, maximum depth 21 metres) lake in northeastern Alberta. The community of Lac La Biche on the lake's southern shore is one of the oldest in Alberta, because it is at a strategic location for the early fur trade. A short portage, the Portage la Biche, connects the Beaver River, which flows into the Churchill River, then on to Hudson's Bay, to Field Lake, which drains to Lac la Biche via Red Deer Brook. In turn, Lac la Biche empties via the La Biche River into the Athabasca River, then northward to Lake Athabasca, the Slave River, Great Slave Lake, the Mackenzie, and the Beaufort Sea. It is one of two historic portages between these two river systems, which allowed voyageurs from York Factory on Hudson Bay to trade for furs with aboriginal people throughout the west and north. Because of Lac la Biche's rich fisheries and convenient travel options, native people undoubtedly inhabited the area for thousands of years.

As a result of fur trade activities, Lac La Biche is one of the earliest European settlements in the province. The first known European settlement was a Northwest Company trading post, Red Deer's Lake House, established by David Thompson in 1798. A Hudson's Bay Company (HBC) post, Greenwich House, the first HBC post beyond Rupert's Land, was established the following year by Peter Fidler. The Portage la Biche was the gateway to the southern Athabasca country from 1799 to 1824 and a passage to the Pacific coast from 1811 to 1824. It was a key to the fur trade that reached from the Atlantic to the Pacific. Greenwich House was closed in 1802, but the HBC returned periodically in an attempt to compete with free traders. However, with the settlement of the prairies, Hudson's Bay governor George Simpson issued a directive abandoning the Beaver River route because it was so shallow and shifting transportation to the recently

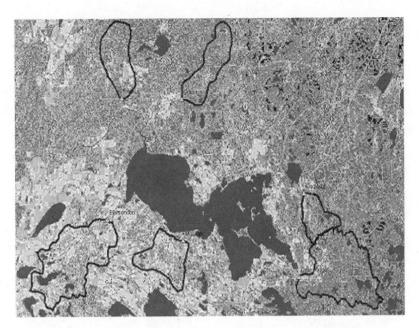

FIGURE 11.2: *Lac la Biche and its catchment. The red and green areas are forested, dominated by conifers and aspen, respectively. The white and gray areas south of the lake are cleared for agriculture. The town is shown as a gray blob in the southernmost tip of the lake. The black lines outline catchments studied for nutrient yield. The main inflow, the Owl River, enters from the northeast, draining largely untouched forests and small lakes. Of the portion of the basin shown, about 50% has been converted to agriculture or urban area. For colour image see p. 333.* Figure from Neufeld (2005).

constructed Fort Edmonton–Fort Assiniboine trail, and later to a trail from Fort Edmonton to Athabasca Landing, upstream on the Athabasca River.

After the closure of the post, Métis settlements remained on the lake, and the Lac la Biche Mission on the shore of the lake was a popular spot. A Cree settlement, the Beaver Lake First Nation, also has remained in the immediate area. There are several other aboriginal and Métis settlements in the region. The mission, now a historic site, still stands on the south shore of the lake.

The earliest impact of humans on the lake was probably to the fishery, which was a staple for both humans and their dogs, kept to provide winter transportation via dogsled. Various reports refer to lake whitefish (*Coregonus clupeaformis*) as "the staff of life" and the lakes as "veritable pantries." The Métis referred to the lakes as "the good Lord's warehouse."

According to the official census of 1872, there were 103 men, 114 women, and 360 children at Lac la Biche, predominantly "French halfbreed." Mission reports later in the 19th century put the human population at 500-600 souls, with the mission still the centre of human activity. According to early photographs, each dog team had four to six dogs. There is a report by Father E. Grouard OMI that there were 400 sled dogs at the mission at Midnight Mass in 1878, competing with the church choir. Various reports indicate that each dog would have received one to three large whitefish per day, probably dependent on how cold it was and how hard they were worked.

Henry J. Moberly reported that the HBC post at Lac la Biche required 9,000-10,000 whitefish each year. The daily allowance was one whitefish per woman, and one half fish per child. In the 1899 Department of Interior annual report, P.R.A. Belanger, DLS, a 19th century surveyor, records over 100,000 whitefish being caught during spawning in two successive years. Concern about the whitefish population led to closure of the lake for a few years in the late 19th century, causing considerable hardship for residents.

In 1915, the Alberta & Great Waterways Railway reached the lake. The centre of activity shifted to the current town site, first called Lac La Biche Station. By 1918, there were four large fish companies operating in the area. Estimated harvests were 500,000 to 1,000,000 pounds of fish per year, as "dressed weight." By the 1920s, the whitefish fishery was in decline, and the fish companies moved on to Lake Athabasca. Commercial fisheries continued to increase, in part to supply several mink ranches developing in the area with food, as cisco (tullibee; *Clupeaformis artedi*). At one time, it is estimated that 75,000 mink were kept near the lake. The greatest catch of cisco was 1,400,000 kilograms in 1946. The cisco population crashed in 1948 because of a massive fish dieoff, perhaps the first sign of increasing eutrophication. After that, the commercial catch began to dwindle. Some commercial fishing continues, and sport fishing increased in the latter part of the 20th century. As a result of overfishing, the walleye (*Sander vitreus*) fishery collapsed (Figure 11.3). The northern pike (*Esox lucius*) fishery has rebounded somewhat from overfishing, thanks to recent more stringent fishing regulations, but it remains in less than optimum condition.

Clearing of the watershed for agriculture began in the late 19th century and accelerated after World War II, as modern tractors became available. In the mid-20th century, when automobiles became reliable and roads improved, the town began to grow. A small (300 people) French-speaking

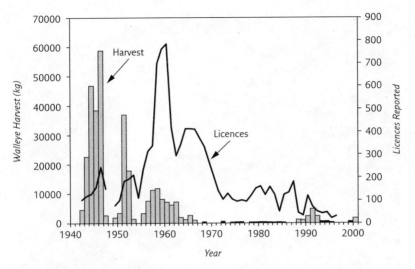

FIGURE 11.3: *The commercial harvest of walleye in Lac la Biche since 1942, showing the current state of collapse. Bars show the harvest, the line shows the number of commercial licences issued by the provincial government.*

Figure from Sullivan (2003). Printed with permission from the American Fisheries Society.

community, Plamondon, the only other community on the lake, is nestled in the valley of Plamondon Creek, which enters Lac la Biche at the western end of the lake.

Recreational summer cottages sprang up around the lake, with early concentrations along the southern and southeastern shores. Lac La Biche was officially declared a town in 1951. A causeway was built to Big Island in the lake to form Sir Winston Churchill Park, opened July 1, 1968. There is considerable local debate as to whether the restricted circulation of the lake caused by the causeway construction is in part responsible for deteriorating water quality.

With steady growth in the human population of northern Alberta and continued improvements in roads and cars, the population around the lake has continued to grow. Much of the southern half of the watershed has also been cleared for agriculture, and logging has begun to make inroads into the basin of the Owl River, the largest tributary to the lake.

Sewage began to be discharged to the lake in 1951, as the town of Lac La Biche constructed its first waste treatment plant. In 1983, the residents, who also drew their drinking water from the lake, decided that this was not such a good idea. Sewage was diverted, discharging to Field Lake,

FIGURE 11.4: *The surface water of Lac la Biche near the mission, September 2005. The "green paint" colour is the result of a massive bloom of nitrogen-fixing Aphanizomenon. For colour image see p. 333.* Photograph by DWS.

FIGURE 11.5: *Dr. Alex Wolfe and Dr. Suzanne Bayley measure a sediment core taken from Lac la Biche, February 2003.* Photograph by DWS.

upstream of Lac la Biche. Subsequent studies showed that much of the sewage drains back to Lac la Biche via Red Deer Brook. Provincial agencies referred to the sewage as "growth enhancing effluent" and regarded Field Lake as of marginal importance. The treatment plant was upgraded in 1989 but continued to discharge to Field Lake. The area around Field Lake was declared a Provincial Historical Resource in 1993, the result of its significance as a historical portage route. A plan for a new treatment plant to remove nutrients or divert sewage from Field Lake is under discussion.

Nearby developments in the pulp and paper industry and in the oil sands caused the community of Lac La Biche to enter a period of rapid growth in the 1990s. A golf course was constructed east of the town. By the late 1990s, the lake was turning very green in summer, and residents were complaining about poor water quality. The lake remains in this condition today (Figure 11.4). Swimmer's itch and high fecal coliform counts are common near some beaches in late summer and have caused some beach closures.

By the 1990s, residents of the town of Lac La Biche and cottagers were increasingly complaining about the intensity and duration of algal blooms on Lac la Biche. In the early 1990s, Alberta Environment did a few studies of the lake and constructed some "strawman" nutrient budgets, based largely on data drawn from other lakes in the area. They concluded that there had been no recent changes to the lake.

Local residents did not believe these results. In 2002, Lakeland County approached DWS to study the lake. Because of the few background data on the lake, a paleolimnology study to determine how much the lake had changed appeared to be a logical starting point. In February of 2003, a group from the University of Alberta, led by DWS and Dr. Alex Wolfe cored the lake through the ice, taking 16 cores from different basins of the lake (Figure 11.5). These were sliced at 0.5 centimetre intervals, and analyzed for nutrients, diatoms, dry weight, algal pigments, and cladoceran remains. Other samples were sent to the University of Ottawa, where they were dated using lead-210.

The cores told a story that supported the concerns of local residents. There was an increase in concentrations of phosphorus, nitrogen, and carbon in sediments deposited during the 20th century (Figure 11.6). Increased inputs of both phosphorus and nitrogen over time caused an increase in algal production. The resulting increase in sedimentation of organic matter caused carbon in sediments to increase as well. Phosphorus increased more rapidly than nitrogen during the 20th century. The declining N:P ratios were toward those known to favour Cyanobacteria, as discussed in Chapter 9 (Figure 11.6).

The slow increases in nutrients are typical of what happens when forests are cut, when land is tilled, and when construction takes place near lakeshores.

The increases in concentrations of nutrients do not tell the entire story. Disturbance in the catchment of the lake causes increased erosion, causing increases in deposition of both inorganic and organic matter to the lake (Figure 11.7). Therefore, the magnitude of the increased nutrient entering the lake per year is the product of the concentration and the increased deposition. This *flux* of nutrient increases more rapidly than concentration (Figure 11.7).

The biological fossils showed exactly the pattern that would be expected from the increase in nutrients. There were increases in the concentrations of fossil remains with time. Biogenic silica, the silica that is deposited in the frustules of diatoms and other silicious fossils like chrysophycean cysts, also increases toward the surface (Figure 11.8). There are also changes in the species composition that reflect increasing eutrophication. Diatom species of the genera *Aulacoseira*, *Cyclotella*, *Fragilaria*, and *Stephanodiscus* that are indicative of moderately productive mesotrophic to eutrophic conditions originally occupied the lake, the result of the phosphorus-rich soils in the lake's catchment. But starting around 1910 in the large basin of the lake (D3), the diatoms began to shift to species typical of more productive lakes. In the later years of the 20th century, populations of some of the original diatom species actually collapsed, at the expense of further increases in *Stephanodiscus niagarae*, *Fragilaria crotonensis*, and other species indicative of highly eutrophic conditions. (Figure 11.9).

Fossil pigments also indicated increasing algal abundance and changes in the composition of the phytoplankton community. Chlorophyll *a* increased, indicating an increase in overall abundance of algae. Increasing fucoxanthin and diatoxanthin confirm the increase in diatoms that we have seen in the hard fossils and biogenic silica. Increasing alloxanthin indicates that chrysophyceans are increasing, confirming the increases in chrysophycean cysts noted above. The increase in several cyanobacterial pigments (zeaxanthin, canthaxanthin, and echinone) show that, while these groups were present in the lake before Europeans arrived, they have increased greatly for the past several decades. In particular, canthaxanthin

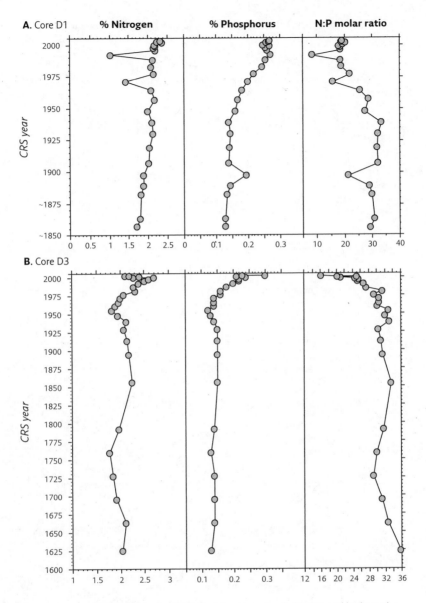

FIGURE 11.6: *Changes in concentrations of phosphorus and nitrogen and in the molar ratio of N:P in Lac la Biche sediments.* From Schindler et al. (2008a).

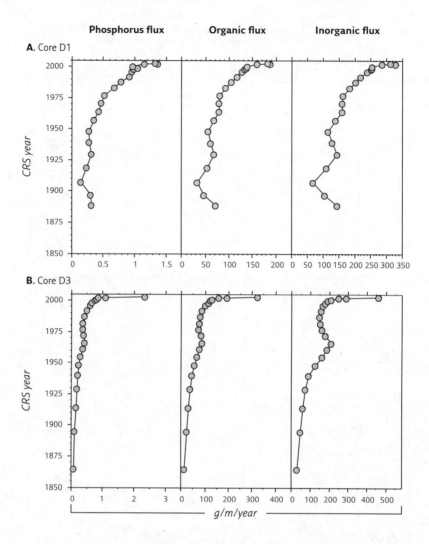

FIGURE 11.7: *Fluxes of phosphorus, inorganic matter, and organic matter to the sediments of Lac la Biche.* From Schindler et al. (2008a).

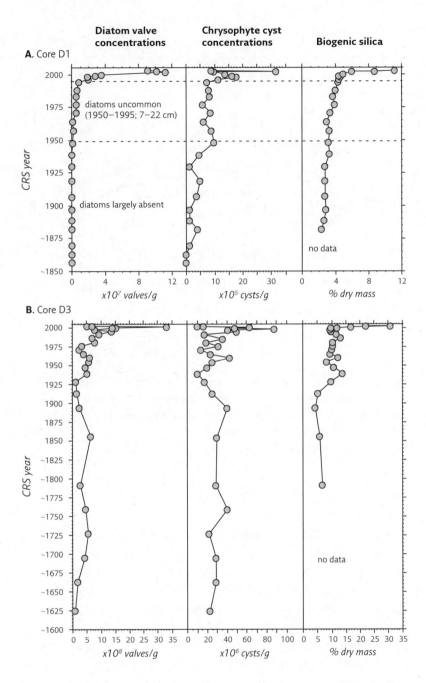

**A.** Core D1

Diatom valve concentrations
x10⁷ valves/g

diatoms uncommon (1950–1995; 7–22 cm)

diatoms largely absent

Chrysophyte cyst concentrations
x10⁵ cysts/g

Biogenic silica
% dry mass

no data

**B.** Core D3

Diatom valve concentrations
x10⁸ valves/g

Chrysophyte cyst concentrations
x10⁶ cysts/g

Biogenic silica
% dry mass

no data

FIGURE 11.8: *The concentrations of diatom valves, chrysophyte cysts and biogenic silica to the sediment of Lac la Biche's eastern (D1) and western (D3) basin.*

From Schindler et al. (2008a).

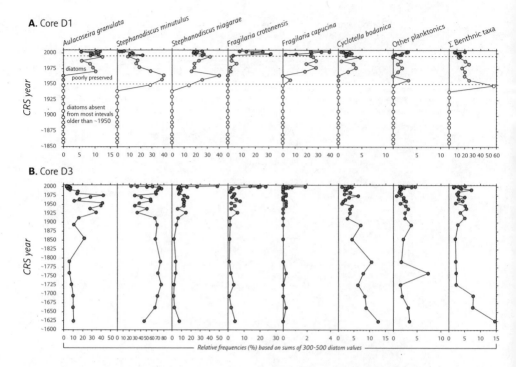

FIGURE 11.9: *Common diatoms in the sediments of Lac la Biche. Note the increase toward the surface of species indicating increasing eutrophy, such as* Stephanodiscus niagarae, Fragilaria crotonensis, *and* Fragilaria capucina. *For reasons not fully understood, preservation of diatoms before 1950 at site D1 was too poor to do long-term interpretations.* From Schindler et al. (2008a).

is indicative of filamentous species of Cyanobacteria, of the sort that now dominate the lake's phytoplankton. (Figure 11.10).

The chemical and biological profiles both confirm what was suspected, that Lac la Biche has become much more eutrophic since the turn of the 20th century, with particularly rapid eutrophication since mid-century when development in the area began to accelerate. The conversion of forests to pasture and farm land has increased the yield of phosphorus per unit area by about twofold (Neufeld 2005). Construction activity at the town site, cottage developments, and road building, including the causeway to Churchill Park, are probably responsible for the increase in silt entering the lake. Some of the phosphorus and nitrogen observed in sediments would accompany soil particles, but sewage from the town's waste treatment centre, septic tanks, fertilizer, manure, and street runoff have increased the

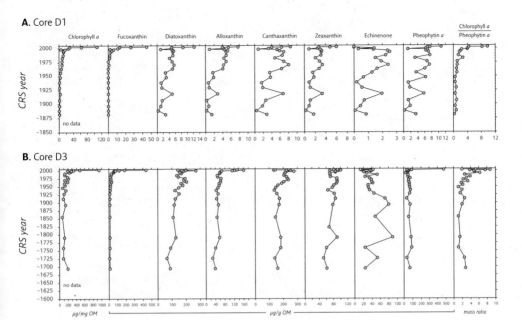

**A. Core D1**

**B. Core D3**

FIGURE 11.10: *Concentrations of algal pigments versus time in sediments of Lac la Biche. Note increase in most toward the surface. OM is organic matter.*
From Schindler et al. (2008a).

concentrations of nutrients as well. The contribution from septic tanks is particularly hard to quantify, but there is a "smoking gun" that suggests it may be important: Jay White of Aquality Consultants has found caffeine in nearshore water samples taken near cottage developments, the result of many cups of coffee passing from human bladders to septic tanks then via groundwater to the lake. Although the leakage of caffeine from septic tanks does not necessarily mean an equivalent leakage of nutrients, it does indicate that water from the septic tanks is reaching the lake. Many studies in other areas have documented the contribution of septic tanks to lake eutrophication. Hopefully, the evidence will help to persuade local residents to replace septic tanks with composting or combustion toilets, and to take other measures to reduce nutrient inputs to the lake. Contemporary studies indicate that sewage, septic tanks, and land runoff all need to be controlled to reduce nutrient inputs to Lac la Biche, if any recovery is to occur.

As described in Chapter 10, a trophic cascade resulting from overfishing the walleye and pike populations may have exacerbated the eutrophication

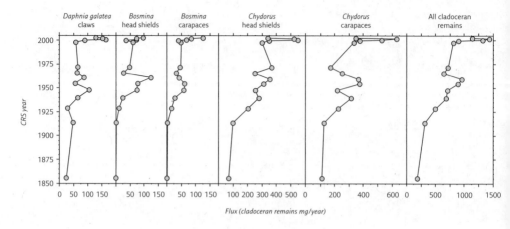

FIGURE 11.11: *Fluxes of the most predominant cladoceran zooplankton* (Daphnia, Bosmina, *and* Chydorus) *in sediment core D3 from Lac la Biche. Note the appearance of* Bosmina, *which is known to feed effectively on Cyanobacteria, early in the 20th century. The increase over time of all three genera is another indicator of increasing productivity in the lake.* From Schindler et al. (2008a).

problem in Lac la Biche as well, by causing the lake to shift from a low-algal to a high-algal phase in the "cascade." Also, preliminary paleoecological evidence suggests a shift in zooplankton, with the cladoceran *Bosmina*, a blue-green specialist, appearing during the 20th century (Figure 11.11).

Some cyanobacterial pigments were present even in the deepest strata sampled, in sediments deposited before 1900 (Figure 11.10). It appears that the lake was mesotrophic to eutrophic even before the rapid growth of human populations in the 20th century, confirming the conclusions made from interpreting changes in diatom species. Therefore, the lake was very susceptible to increasing nutrient inputs. Anoxic conditions occur in deeper strata under winter ice and during short periods in summer (typically no more than a few days) when the lake is calm. During these calm periods, oxygen is rapidly depleted and release of phosphorus from sediments is high. When ice leaves in the spring, and when wind following calm periods circulates the water of the lake in summer, these nutrients are swept into near-surface water to replenish algal blooms. If nutrient sources are reduced and fisheries restored, Lac la Biche may recover. However, the recovery will probably be slow, expensive, and difficult. It will take many years for sediment sources of phosphorus to be depleted or buried deeply enough to minimize their exchange with overlying water. Fisheries scientists indicate that the restoration of the sport fishery may require 20 years or more. They propose that it would require a reduction in nutrient inputs, an aggressive campaign to reduce cormorants, and stocking with millions of walleye fingerlings.

Lac la Biche is not alone as a lake that has undergone eutrophication in the 20th century. Throughout southern Canada, lakes are literally being "loved to death" by people who do not understand the combined effects of all their activities on the nutrient balances of lakes. Similar stories are told by the sediments of Lake Winnipeg, the Qu'Appelle lakes of Saskatchewan, and many other lakes and reservoirs of the prairies, as well as those of Muskoka, Minnesota, and other areas where development is occurring near lakes. The Algal Bowl is thriving, especially in the West!

The above is but one example of how paleolimnology can be used to determine the causes of historical changes in a lake or its catchment. For a more thorough treatment, see Smol (2008).

# 12 | Recovery from Eutrophication

*The crude methods that are currently available*
*for recovering culturally eutrophied lakes*

"We end, I think, at what might be called the standard paradox of the
20th century: our tools are better than we are, and grow better faster
than we do. They suffice to crack the atom, to command the tides. But
they do not suffice for the oldest task in human history: to live on a piece
of land without spoiling it." —ALDO LEOPOLD

IN EARLIER CHAPTERS, we gave several examples of successful recoveries of lakes after phosphorus inputs were reduced. However, not all lakes recover rapidly after nutrient loading is reduced. In many lakes, phosphorus return from lake sediments to the overlying water can be the largest source of phosphorus to the water column in midsummer. Of course, this phosphorus has originally come from outside the lake, but as the result of saturating the top several centimetres of sediments, it can return to the water column for many years, either by diffusion or by the resuspension of surface sediment.

If a lake is eutrophic for many years, many centimetres of phosphorus-rich sediments can accumulate on a lake's bottom. Normally, after external loading of phosphorus is decreased, it takes several years for the phosphorus-rich sediments to be buried by low-phosphorus sediments. Often, this is not fast enough to satisfy a public impatiently expecting near-instant recovery once they have controlled nutrient resources. The result has been an explosion in contraptions and techniques that are supposed to help lakes recover, including aerators, chemicals for preventing sediments from releasing phosphorus, and even blue dyes that are poured into lakes to mask the green colour caused by algal blooms!

Physical, biological, and chemical remediation techniques are some-times used. Physical techniques include dredging to remove organic sediments that are rich in phosphorus and consume large amounts of oxygen, or artificial destratification, which prevents the formation of stag-nant hypolimnions that can become anoxic. Biological techniques can include manipulation of fish stocks to flip the "trophic cascade" from a high-algal phase to a low-algal phase, as described in Chapter 9. In other cases, macrophytes have been reintroduced, to attempt to change the lake from a turbid water to a clearwater state. Chemical techniques include injection of nitrate into bottom sediments, where denitrifying bacteria release oxygen, preventing reducing conditions that release phosphorus. Still another is to add enough alum or iron to deepwater to precipitate complexes of phosphorus with aluminum or iron. Finally, addition of oxygen to the hypolimnions of lakes has been used to prevent the anoxic conditions that favour phosphorus release. All of these measures have been successfully used alone or in various combinations in small to medium-sized lakes or ponds, but none can claim total success (Cooke et al. 2005). They are all too expensive to be feasible in large lakes, *where the most cost-effective approach is still to reduce the external nutrient load and wait*, often for years or even decades for the symptoms of eutrophication to subside.

Recovery of lakes from eutrophication can be an expensive and time-consuming business, and it is prudent and cost-effective to prevent eutrophication in the first place. Unfortunately, this is a hard lesson for people to learn. There are now thousands of case histories of eutrophied lakes and expensive partial restorations, differing only in minor detail. However, people always hope that their lake will react differently, somehow allowing them to profitably develop and exploit the catchment while magically avoiding the well-known effects of adding nutrients. The results are always the same, declining transparency, floating algal blooms that decay on the shores, and finally the decision to do something to reverse the problem...rather like closing the barn door after the mare's been bred! Here, we give a few case histories, to show readers what the result can be of ignoring eutrophication in the race to profit from "development."

When the residents of lakeshores and catchments finally take action, they expect the lakes to respond instantaneously; however, when lakes have been eutrophic for decades, this is extremely unlikely. For recovery to take place, the phosphorus-rich surface sediments, which have been accumulating for years, must often be buried under enough new, phosphorus-poor sediment to prevent phosphorus from escaping back to the lake water. This can take many years. In a plenary lecture to the 29th Congress of the International Society for Theoretical and Applied Limnology in Lahti, Finland, in August 2004, Dr. Erik Jeppesen of Denmark summarized recovery efforts for a large number of shallow Danish lakes. He found that, once external loading was reduced, lakes often required 10 years or more for phytoplankton to respond fully to decreased phosphorus inputs. In some cases, no significant recovery was noted even 30 years after nutrient inputs were reduced. Steve Carpenter used Wisconsin lakes as the basis for a predictive model to estimate the time that lakes would require for full recovery. The factor that slowed recovery the most was catchment soils that had been saturated with phosphorus by decades of heavy fertilizer application. The residence time of phosphorus in soils was hundreds of years, much longer than even in lake sediments (Carpenter 2005). In short, great patience is required to recover many lakes from eutrophication, and it is advisable not to allow lakes to become eutrophic in the first place. The old adage "a gram of prevention is worth a kilogram of cure" fits eutrophication perfectly.

Carpenter's models suggest that in the original, pristine condition where the catchment of a lake was undisturbed, had low phosphorus in runoff, and a had healthy aquatic community normally keep a lake in a clear-water phase. As the phosphorus loading increases, there are two possible phases: a clear-water phase, where most of the increased nutrients are taken up by aquatic macrophytes, or a turbid water phase, where algal growth shades out macrophyte growth. Some lakes fluctuate between these two states at unpredictable intervals, with a switch triggered by some sort of event (like re-suspension of silt and sedimentary phosphorus by wind) or a change to the community (changes in the "grazing power" of the herbivore community, usually because of changes to the fish communities that control the "trophic cascade" as explained in Chapter 9). Lake managers try to manage such ecosystems to maximize the time that the lake is in the clear-water phase, because the lake is more aesthetically pleasing and because the water is more suitable for drinking or swimming. The macrophytes also provide habitats for fish, so that the clear-water phase has higher fish production. Waterfowl, too, appear to prefer the clear-water phase.

As phosphorus loading increases, the turbid state becomes more and more predominant. Eventually, a lake can become permanently turbid, and macrophytes are shaded out. If this condition persists for years, macrophyte seedbeds can be buried or destroyed, making it impossible to "flip" the system back to a clear-water state, even with greatly reduced phosphorus loading. Such lakes are difficult to restore. Some lakes may even be impossible to recover from eutrophication. It is preferable to reverse eutrophication before such a new "stable state" is reached.

## Some Case Histories of Recovery Attempts

THERE ARE EXAMPLES where humans have "pulled out all the stops" to attempt to restore lakes more quickly. Many of the examples are in Europe, where much more attention has been given to the restoration of eutrophic lakes than in North America. One excellent example was the subject of a midweek excursion at the above-mentioned limnology congress. The following account is taken from the presentation on that excursion.

Tuusulanjärvi (Lake Tuusula) is a medium-sized (area 5.9 km², maximum depth 9.8 m) lake north of Helsinki, Finland. Development of the 92 km² catchment of the lake began to accelerate early in the 20th century, when the lake became a fashionable summer retreat for city people. Increasing areas were also cleared for agriculture. The lake quickly became a cultural centre and home for many famous Finnish artists, writers, and musicians. The best known to westerners is probably composer Jean Sibelius, who fell in love with the natural beauty of the lake and its surroundings. His home, Ainola, overlooking the lake from the eastern side, was completed in 1904. Sibelius claimed that the silence of the lake was necessary for him to compose. His composition was done entirely in his head, without the aid of a piano or other instrument. He believed that noise interfered with the purity of tone in his mind, and he demanded total silence while he was composing. When he was given a piano as a gift, he demanded that his daughters only play it when he was out of the house. Today, the rumble of traffic would probably prevent Sibelius from composing altogether! One wonders how many equivalent maestros in music or other creative occupations are foregone today by simply not having a quiet environment in which to allow their minds to unleash their creative powers!

For the next several decades, land clearing, draining of wetlands, and development proceeded in the catchment of Lake Tuusula. Roads were built to ease access for tourists, and several small communities based on tourism joined artists' communities, museums, and an armed forces base near the lake's shore. The city of Järvenpää developed at the northern edge of the lake, using the lake as a repository for its sewage. Land cleared for fields was fertilized for agriculture, first with manure, then with artificial fertilizer. Already in 1937, the famous Finnish limnologist Dr. H. Järnefelt had noted that algae were increasing in the lake. By mid-century, few of the original forests of the basin remained. A dam at the outlet raised the water level, and reduced water renewal. Algal blooms intensified even more. People shunned the lake, which became known for its huge floating blue-green blooms and awful stench, rather than its pristine beauty. Secchi depth decreased to a few tenths of a metre, and chlorophyll *a* concentrations reached nearly 100 micrograms per litre at the height of summer blooms. Anoxic conditions under winter ice caused fish kills.

In the 1970s, residents had had enough. A Water Protection Association began to undertake efforts to restore the lake to its earlier condition. Initial efforts to restore the lake were limited to aeration under winter ice. Although these reduced fish kills, they did little to reduce summer algal blooms. In 1979, municipal wastewater from Järvenpää was redirected away from the lake, removing about half of the estimated average of 10 tonnes of phosphorus that the lake received each year. The loading since that time has averaged about 4.6 tonnes per year, all from non-point sources, including 1700 homes still on septic tanks, agricultural fertilizers, golf courses, and street runoff. Efforts are underway to reduce these sources of nutrient pollution. Compensation has been offered to farmers to reduce nutrients from agricultural runoff that reach the lake via huge drainage ditches.

Recovery has been slow, with considerable year-to-year variation. In 1992, cleaner water from a nearby lake was pumped into Lake Tuusula. It had little effect. In 1997, following the largest algal blooms in a decade, more intensive measures were implemented. Cyprinid fish (roach, *Rutilus rutilius*) were removed from the lake, hoping to reverse a "trophic cascade" set off by overfishing for piscivorous predators. As of 2004, 422 tonnes of fish had been removed, equivalent to 3000 kilograms of phosphorus. More aerators (six as of 1998) were installed, operating both summer and winter, keeping oxygen concentrations high most of the time in the small hypolimnion. However, bottom waters still go anoxic during calm, warm periods, despite the pumping of 460,000 cubic metres of oxygen-rich epilimnion water per day into the hypolimnion. Wetlands were built in 2001 to receive the runoff from Sarsalanova Brook, one of the larges sources of nutrients. While the wetlands are still maturing, they remove only 15% of the phosphorus from this source.

There are plans to remove 1,000,000 cubic metres of phosphorus-rich sediments from 1.2 square kilometres of the deepest part of the lake. The removed sediment would be transported to a constructed sedimentation basin of 150 hectares. The cost would be 3.2 million Euros (4.2 million U.S. dollars). An alternative scheme is also being considered. This would involve spreading ferric sulfate over the sediments to prevent phosphorus from escaping, followed by adding a 10 cm layer of clean clay. This option alone is estimated to cost 1.7 million Euros (2.2 million U.S. dollars). There is also the problem of storm water and street runoff from cities and developments in the watershed of the lake, which will be costly to correct.

The total of all of these expensive supplementary measures has resulted in the reduction of blue-greens from 50%–90% of algal biomass in the 1970s to 20%–50% in the early 21st century. The lake still had transparency less than 1 metre in August 2004, when DWS visited the lake with a conference tour. Chlorophyll *a* has decreased below the mean trajectory of earlier years but not by much. It is estimated that, overall, external loading must be reduced even more, to about half of the current 5 tonnes of phosphorus per year for the overall plan to be successful. Altogether, it is apparent that by far the most cost-effective measure to restore a eutrophic lake is to reduce external loading and be prepared to wait, perhaps for more than a generation. Certainly, the expensive and massive restoration efforts on Lake Tuusula should give pause to societies that wish to risk excessive development near lakeshores and the streams that enter lakes.

Danish scientists also have extensive experience with lake restoration (Søndergaard et al. 2000; Jeppesen et al. 2005). They have used a variety of methods in over 35 lakes, studying the results for well over a decade. They advised that external loading must be reduced to get phosphorus in a lake below 0.05–0.1 milligrams per litre before it is worth applying other methods to reduce internal loading. The most successful secondary methods involved manipulation of the "trophic cascade" to maximize the grazing power of zooplankton. They estimated that 80% of the zooplanktivorous fish would have to be removed to have a positive effect on lake recovery. To keep zooplanktivorous fish at this low level usually requires the establishment of healthy populations of piscivorous fishes. Hypolimnetic oxygenation with air, adding oxygen or nitrate to surface sediments, dredging, and even the replanting of destroyed macrophyte beds were also tried.

Even with all of these methods to apply, not all Danish lakes have recovered. Lake Nordborg, on the island of Als, received the sewage from the town of Nordborg, which covers 25% of the lake's catchment. Also, 63% of the catchment is used for intensive agriculture. Diversion of sewage from the lake in 1988, followed by other measures, reduced the phosphorus loading from 1600 to 500 kilograms per year by 2002. However, the internal loading from sediments remains at 1300 kilograms per year in summer. New initiatives are planned, including biomanipulation, alum treatment, and further controls on surface runoff.

Swedish scientists have monitored the recovery of the four largest lakes in Sweden for over 40 years. Phosphorus inputs were reduced significantly in the mid-1970s. Recovery has been slow but steady since that time (Willen 2001; Wilander and Persson 2001).

Lake Lucerne, Switzerland, has been studied since 1961. In its pristine state, it was a deep, oligotrophic lake in the Alps. In the 1960s and 1970s, the lake eutrophied rapidly. Total phosphorus concentrations increased sixfold, Secchi depths decreased from over 4.5 metres to less than 3.5 metres, with *Planktothrix* abundant in the phytoplankton. After phosphorus inputs were reduced in the late 1970s, the lake has slowly recovered. The lake's productivity and nutrient concentrations in the late 1990s were similar to those 50 years earlier. However, the species of phytoplankton were different (Bührer and Ambühl 2001; Bossard and Bürgi 2005). It is believed that changes to the fish community during the past half century were important in restructuring the lake's community.

Lake Geneva, another huge alpine lake, on the border between France and Switzerland, became increasingly eutrophic during the 1970s. Measures to reduce phosphorus inputs began in 1981. By the late 1990s, phosphorus concentrations in the lake had declined to about half of those when the lake was most eutrophic. To the dismay of lake managers, the phytoplankton in the lake have not declined at all. They are, however, quite low by the normal standards of eutrophication, 5–8 micrograms per litre as chlorophyll *a* (Anneville and Pelletier 2000).

Shallow lakes in the Netherlands also became eutrophic in the mid-20th century. Dense algal blooms shaded out macrophytes in the lakes, switching them to domination by algae. The lakes recovered more slowly than they became eutrophic, with algal blooms reluctant to disappear. Manipulation of fish communities has been somewhat successful, but macrophytes have been slow to recover. As in the examples above, recovery times are many years. In some cases, the systems appear to have shifted to alternative stable states, rendering recovery extremely difficult and perhaps impossible (Scheffer et al. 2001).

Many more examples could be described, but the above should suffice to illustrate that successfully recovering eutrophic lakes is problematic at best. It is clear that prevention should be the main strategy for controlling eutrophication, if long-term, costly recoveries are to be avoided. The most difficult lakes to recover appear to be so-called polymictic lakes, in which thermal stratification forms and is broken several times during the ice-free season. If such lakes become eutrophic, hypolimnetic oxygen depletion and phosphorus release in deepwater occurs during every stratification event, then stratification is broken by the wind allowing the phosphorus to be re-mixed into the euphotic zone several times each summer (Schindler and Comita 1972). As a result, algae always have plenty of phosphorus, and they grow vigorously all summer. In normal dimictic lakes, a thermocline forms as lakes warm in the spring, and the hypolimnion is not again mixed with surface water until autumn. Phosphorus, once sedimented by dying plankton, remains below the thermocline until fall overturn. In monomictic lakes, which do not stratify significantly at all, the stratification necessary to allow deepwater oxygen depletion does not occur.

Another characteristic of lakes with high return of phosphorus from sediments is that iron always seems to be scarce. When iron is plentiful, iron hydroxides and minerals that contain both phosphorus and iron-like vivianite seem to reduce the internal loading of phosphorus at least some-what. The mechanisms need further investigation.

# 13 | Eutrophication in Estuaries and Coastal Zones

*The expansion of the Algal Bowl to the coastal oceans*

So FAR, we have restricted our discussion of eutrophication to freshwater ecosystems, but there is also great concern about the eutrophication of coastal waters, which was just beginning to emerge at the time of JRV's first edition of *The Algal Bowl*. Today, almost 50% of the world's population lives within 100 kilometres of the seacoast, and the numbers are still increasing. The result is widespread eutrophication.

Most coastal rivers and bays in the United States are moderately to severely degraded. An extensive anoxic "dead zone" in the Gulf of Mexico has been documented off the Mississippi Delta, the result of high nutrients delivered by the river. In addition, dozens of cases of coastal zone degradation by high nutrient loading have been documented in the United States (see, for example, papers in *Limnology and Oceanography*, volume 51(1), 2006). On average, U.S. estuaries receive many times more nitrogen and phosphorus today than they did before Europeans arrived. Agriculture, urban and residential expansion, and land-use changes are all implicated.

So far, Canadian estuaries have mostly escaped eutrophication. This is largely because of strong currents and tides in the few areas that have significant human populations. However, coastal British Columbia is of some concern, because of the rapidly increasing populations in the Vancouver and Victoria areas. They, the city of Seattle, Washington, and many smaller population centres discharge sewage into the same water-body, the Georgia Basin. Altogether, over 4.5 million people discharge their wastes to these waters. The area's population is expected to grow by 28% by 2010 (Schindler et al. 2006). Much of the sewage discharged from Canada to the Georgia Basin receives no treatment at all.

In Europe, the Baltic Sea, the North Sea, the Black Sea, and the Adriatic Sea have been the focus of concerns about eutrophication (Vollenweider et al. 1992). In Asia, severe eutrophication has been observed off coastal Japan, the Yellow Sea, the East China Sea, and many other areas. Both the North Sea and the Yellow Sea now receive 10–15 times the nitrogen inputs that they did in pre-industrial times.

In South America, there have been huge migrations of humans from rural areas into large coastal cities. Six cities in South America now have populations over 10 million. Sao Paulo, Brazil, has almost double that number: 18 million at the last census. Only 10% of its sewage is treated before discharge. Much of the population in these megacities lives in slums that have no water or sewer systems. In short, coastal waters are showing severe signs of eutrophication in most populous parts of the world.

Rapid growth in agriculture has caused some of the eutrophication of coastal waters. For nitrogen, atmospheric deposition in regions downwind of major industries is also important. In most tropical and subtropical countries, use of fertilizer is increasing very rapidly as the need for food increases with rapid population growth. In other cases, much of the food production is for export. For example, in Central and South America, the two crops that are expanding most rapidly are sugar cane and soybeans. The former has doubled in acreage since 1960. Soybeans have gone from almost zero to over 35 million hectares in the same period. It is predicted that these crops will continue to expand, as they are increasingly used to produce ethanol and biodiesel in an attempt to supplement dwindling petroleum reserves, as well as to provide sugar, cooking oil, and food for livestock.

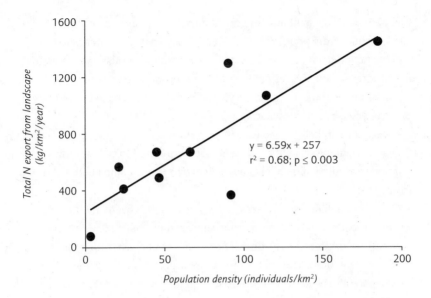

FIGURE 13.1: *The relationship between human population density and export of nitrogen from several river basins. Note that there is some nitrogen export even from catchments that have very low human populations, such as northern Canada.*

From Howarth (1998). Printed with permission from Springer Science and Business Media.

Livestock culture in near-coastal areas also contributes nutrients. In particular, huge confined feeding operations of the sort that we described in Chapter 10 also occur in coastal areas.

Although large human and animal populations in coastal areas are a major cause of the increased eutrophication, the sources of the nutrients for estuaries that drain large watersheds can also be far inland. For example, much of the nitrogen discharged into the Gulf of Mexico by the Mississippi River originates in the fertile farmlands of the central plains, several hundred kilometres upstream. It is estimated to be five to six times the natural loading from the river.

There is a linear relationship between application of new nitrogen as fertilizer and nitrogen loss to river systems. The nitrogen export from watersheds is roughly proportional to human population. Losses in near-pristine watersheds approach 100 kilograms per square kilometre, which is regarded as approximately the natural background for many areas (Figure 13.1). Of regions that have been studied in North America, only Hudson Bay and the Labrador coast still have near-background nutrient inputs, largely because there are small human populations, no agriculture, and sufficient distance from populous areas that atmospheric deposition is low. Air masses polluted with nitrogen oxides and ammonia can also carry nitrogen for hundreds of kilometres before it falls on coastal waters as nitrate or ammonium. In some areas like the Baltic Sea, these can be major sources of nitrogen, as we discuss below.

As a result of the later increase in concern about estuarine eutrophication and the high variability in waters that are referred to as "estuaries," our understanding of managing eutrophication is less complete than for freshwaters.

Much of the focus on estuarine eutrophication has been on nitrogen, because most coastal marine systems show symptoms of nitrogen limitation. Also, nitrogen losses from land are increasing faster than for any other nutrient, or for that matter, any other aspect of global change.

We have already described the confusion in freshwater policies that resulted in the 1970s from the fact that most eutrophic lakes showed clear signs of being nitrogen or carbon limited when small-scale bioassays were used. We were able to demonstrate in lakes 226 and 227 that carbon and nitrogen limitation were caused by fertilization with phosphorus. Clearly, nitrogen and carbon limitation were not indications that control of these nutrients would reduce eutrophication, as had been argued by many limnologists in the 1960s. In short, whole-lake experiments showed that nitrogen and carbon limitation were symptoms, not causes of eutrophication.

It is possible that the same mistakes in interpreting physiological symptoms of nutrient levels are being made with respect to estuaries. Many of the same arguments that confused limnologists and policy makers half a century ago are occurring today in coastal ecosystems. There is still confusion over whether nitrogen limitation means that nitrogen, rather than or as well as phosphorus, should be managed to reduce algal blooms. Although the technology for removing nitrogen from sewage has improved dramatically, it is still expensive compared with removing only phosphorus. Unfortunately, whole-ecosystem experiments in estuaries are rare, most were done before methods for modern water chemistry and biology were developed. Later, we describe an exception: an "experiment" carried out in the Stockholm Archipelago as a result of phosphorus management to protect freshwaters in Sweden.

In addition to their large size, estuaries have an additional complication in being transitional between freshwater and marine systems. Salinity in different estuaries ranges from a few parts per thousand to almost full-strength seawater. At the freshwater end of the gradient, many of the species of algae and other organisms are the same as those in freshwaters. At the saline end, they resemble the communities of the oceans.

There are also complex physical and biogeochemical phenomena at play. Some estuaries can have invasions of higher salinity water from the oceans as the result of tides and winds. The denser saline inputs tend to slide into an estuary along the bottom, causing stratification gradients called *haloclines* that are barriers for the exchange of dissolved nutrients and oxygen. By restricting exchange of oxygen between deep and near-surface waters while allowing plankton and other particles to settle through the halocline and decompose, haloclines often aggravate the anoxic conditions that promote high recycling of phosphorus from sediments, as we describe later.

Nitrogen-fixing Cyanobacteria are less common in high-salinity marine systems. They are reputed to do poorly when salinity exceeds 10–12 grams per kilogram, i.e., about one-third strength seawater. They are increasing in many coastal ecosystems nevertheless. Nitrogen-fixing Cyanobacteria thrive in some inland lakes with high salinities, suggesting that the reason for their scarcity in the oceans is not directly the result of high salinity. Roxanne Marino and her colleagues (2006; see also Chan et al. 2006) found that N-fixing Cyanobacteria thrived in high-salinity (27 grams per kilogram) water from Narragansett Bay, Rhode Island, if zooplankton populations were kept low. If normal concentrations of zooplanktom were present, their grazing kept blue-green filaments too small to form heterocysts, the cells that have the ability to fix atmospheric nitrogen. When zooplankton were low, filaments grew in length, heterocysts formed, and nitrogen fixation occurred. This result is curiously similar to the situation in Lake 227, where as we discussed in Chapter 9, high populations of grazing zooplankton in Lake 227 followed addition of pike (*Esox lucius*) to the lake. Their grazing kept nitrogen-fixing Cyanobacteria populations low, in contrast to their abundance when grazing power was low, in the absence of predatory fish. Perhaps in both estuaries and freshwater, lack of "grazing power" by zooplankton is one factor that allows nitrogen fixers to predominate. As in freshwater, changes have happened to upper levels of the food chains of coastal waters, as well as increased nutrient loading. Also, denitrification appears to be more important in estuaries. Howarth et al. (1999) also note that high sulfate in estuaries limits the availability of molybdenum, a micronutrient essential for nitrogen fixation by Cyanobacteria. Again, many freshwater lakes with Cyanobacteria have sulfate concentrations just as high. It is hypothesized that this micronutrient limitation slows the growth rates of nitrogen fixers, making them more vulnerable to grazing. Overall, management of eutrophication in estuaries appears to be more complex than it is in freshwaters. It has been hypothesized that scarcity of iron could further limit blue-green algae in marine systems. Clearly, we still do not understand the responses of coastal systems as well as we do freshwaters.

## The Baltic Sea

WE SHALL USE the Baltic as an example to illustrate the complexity of the problems of interpreting eutrophication in estuaries. With the possible exception of Chesapeake Bay, the Baltic is the best studied estuarine system in the world. There are some data from over 100 years ago. The Baltic has an enormous salinity gradient, ranging from very fresh water (salinity less than 2 grams per kilogram) in Bothnian Bay and the Bothnian Sea, to nearly full-strength sea water in the outer regions near the North Sea, the Skagerrak, and Kattegat (which some would claim are not really part of the Baltic; Figure 13.2). As we shall see, the Stockholm Archipelago, an important part of the Baltic, was also the site of a huge-scale nutrient-management experiment of the sort that we believe provide irrefutable evidence for the actions needed to control eutrophication.

At present, 16 million people live in nine countries along the coast of the Baltic Sea (Figure 13.2). A total of 85 million people live in the 14 countries in its catchment (Wulff et al. 2001). The distribution of population varies widely and is generally much higher in the southern part of the drainage basin.

Changes to the Baltic Sea have been caused by a variety of insults, including toxic pollutants, overfishing, and climate warming, but eutrophication has been one of the most important stressors (Figure 13.3). Nitrogen inputs to the Baltic have increased fourfold and phosphorus inputs eightfold since the mid-19th century (Larsson et al. 1985). Although paleoecological studies show that Cyanobacteria were always present in phytoplankton of the Baltic, they have increased both in importance and relative abundance in the 20th century. The objectionable

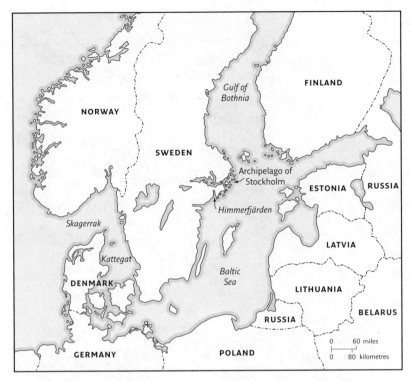

FIGURE 13.2: *Map of the Baltic Sea and much of its surrounding watershed.*
Map by Wendy Johnson.

symptoms of eutrophication vary widely in different parts of the Baltic. They include massive algal blooms of Cyanobacteria (chiefly *Nodularia* and *Aphanizomenon*) and anoxic "dead zones" in bottom waters, including the large, deep basins of the Baltic. In total, the dead zones of the Baltic are estimated to exceed 70,000 square kilometres. In these dead zones, the mixing of sediments has ceased because bottom-living organisms were extirpated by lack of oxygen. The result has been that sediments are now laminated because of not being mixed by benthic organisms, as discussed below. In the shallower nearshore fjords and archipelagos, development of littoral mats of filamentous green algae, disappearance of the seaweed *Fucus* and eelgrass beds, and lower transparency are frequent objectionable symptoms of eutrophication.

In response to frequent concerns about eutrophication, some Baltic countries began to reduce point sources of phosphorus in the early 1970s. This was driven largely by evidence from freshwater systems and partly

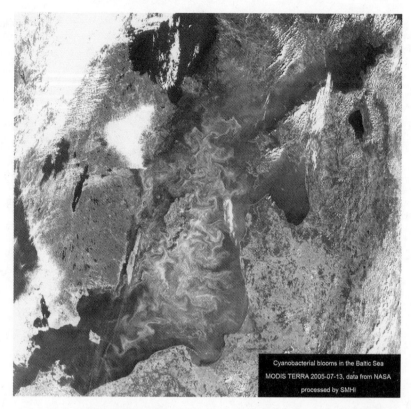

Cyanobacterial blooms in the Baltic Sea
MODIS TERRA 2005-07-13, data from NASA
processed by SMHI

FIGURE 13.3: *Cyanobacterial bloom in the Baltic Sea, July 13, 2005. The image is from the satellite sensor MODIS (Moderate Resolution Imaging Spectroradiometer). For colour image see p.334.* Satellite data is from NASA, GES Distributed Active Archive Center, and the data are processed by the SMHI.

by the relative ease of removing phosphorus from domestic sewage. As described below, the measures were very successful in the waters of the inner and middle Stockholm Archipelago. However, despite significant reductions in nutrient inputs, there was little visible effect on the Baltic Proper or on the Kattegat.

Atmospheric nitrogen deposition has been another problem for the Baltic and in other estuaries surrounded by human activity. As a result of high emissions of nitrogen oxides and ammonia by automobiles, industries, and agriculture in the airshed of the Baltic, nitrogen deposition increased until about 1980. Deposition leveled off in most countries after that and declined in some (Tarranson and Schaug 2000). Atmospheric sources continue to be a large input of nitrogen to the Baltic, with values in the Baltic Proper exceeding 14 kilograms of nitrogen per hectare per year (Granat 2001). High atmospheric deposition in terrestrial parts of the Baltic catchment contributes to the riverine loading as well. Wright et al. (2001) found that 10 kilograms of nitrogen per hectare per year was a deposition threshold, above which nitrate in stream runoff could be expected to increase above background levels, indicating that terrestrial ecosystems were nitrogen saturated. Although there have been some attempts to control emissions of nitrogen to the atmosphere in Europe, Skjelkvåle et al. (2005) found no evidence of declining concentrations of nitrate in surface waters, including those in the Baltic region.

Atmospheric deposition of nitrogen has also caused detrimental changes to freshwaters and forest soils of many regions around the Baltic (Dise and Wright 1995). The precursor nitrogen oxides are an important component of smog. For these reasons, atmospheric sources of nitrogen should be controlled, regardless of their effect on eutrophication.

Despite significant reductions in nutrient inputs from point sources, eutrophication in the Baltic has continued to worsen. In less saline parts of the Baltic, including the Baltic proper, nitrogen fixation by Cyanobacteria adds significantly to nitrogen inputs in midsummer. Fixation does not appear to significantly enhance nitrogen inputs to the west coast and Kattegat, where salinity is thought to be too high for Cyanobacteria blooms to develop.

Increasing temperatures have also occurred in the Baltic (Fonselius and Valderrama 2003). These have probably caused both production and decomposition rates to increase, compounding the effects of nutrient loading.

Different results from different regions of the Baltic, different interpretations of results and a shortage of definitive data have caused lively and at times acrimonious debate among scientists over what measures are necessary to reverse eutrophication. Most scientists agree that phosphorus must be controlled, but some believe that reversal of eutrophication will be more rapid if both nitrogen and phosphorus are controlled. Still others believe that controlling nitrogen is a waste of money and that the focus of reductions should be on phosphorus alone, especially because in many Baltic countries there is no control of nutrient discharges at all. Some argue that reducing nitrogen could increasingly favour nitrogen-fixing blue-green algae, which are largely the focus of current concerns by those who use the Baltic for recreation.

In 2005, DWS was invited to be a member of a five-person international review panel to advise the Swedish government on the matter of controlling nutrient loading to the Baltic, especially with respect to Swedish sources of nutrients. The panel ended effectively in deadlock. All members strongly recommended that phosphorus must be controlled, but two members believed that nitrogen control was essential as well. Two panel members (including DWS) believed that, given the realities of available funds, the entire emphasis should be on external phosphorus sources. One panel member was "on the fence." The following is largely abbreviated from that panel's report, but interpretations are biased by the viewpoint of DWS.

### The Stockholm Archipelago

THE STOCKHOLM ARCHIPELAGO contains at least 20,000 islands. It is one of the most important recreational areas for Sweden. In summer, the inlets and bays are alive with sailboats and motorboats. Regular and frequent ferry service transfers crowds of people to and from the mainland to summer homes. Some even commute daily from the islands into Stockholm. A number of resorts, some luxurious, others rustic, accommodate tourists from international destinations.

Land cover in the drainage basin of the archipelago is a mixture of agriculture, urban areas, and forest. The surface waters are brackish, with salinities generally less than 6 grams per kilogram, with higher salinities in bottom waters. The dominant freshwater input is from Lake Mälaren, with a mean out-flow of approximately 165 cubic metres per second. Water residence times in the archipelago range from a few days in the outer region to over 100 days in the inner archipelago. The archipelago receives nutrient inputs from sewage treatment plants serving the Stockholm metropolitan area, and from the surrounding watershed through the outlet of Lake Mälaren, the so-called Nörrstrom outflow. Discharges to the archipelago from the three largest sewage treatment plants (STPs), Henriksdal, Käppala, and Bromma, are at depths of 30–45 metres, below the summer thermocline and winter halocline. The numerous permanent and seasonal homes on the shores and islands of the Archipelago are also potential sources of nutrients. Most are served by septic tanks. Perhaps significantly, Sweden never banished the use of phosphorus-based detergents, choosing instead to remove the phosphorus from sewage discharge. However, if wash water goes to septic systems, the phosphorus escapes removal, unless it binds to soils. In many cases, soils become saturated, after which phosphorus can seep into water.

As early as 1900, the wastewater from approximately 200,000 people discharged into the Stockholm Archipelago (Brattberg 1986). In 1941, the largest treatment plant in Stockholm (the Henriksdal) began operation, with wastewater discharges of nutrients into the archipelago increasing markedly until about 1970. At that time, about 70% of the phosphorus inputs to the archipelago were from STPs and about 30% from the Lake Mälaren outflow. In contrast, only about 40% of nitrogen inputs were from STPs and about 60% from the Lake Mälaren outflow.

From the 1940s to the late 1960s, phosphorus inputs increased more rapidly than nitrogen (Figure 13.4). As a result, the ratio of total nitrogen to total phosphorus decreased to about 7:1 by weight (Table 13.1).

*Table 13.1: Total nitrogen to phosphorus (N:P, by weight) of inputs to the Stockholm Archipelago.*

| Period | N:P |
|---|---|
| Late 1960s | ~7:1 |
| 1980–1985 | 32:1 |
| 1990–1995 | 38:1 |
| 1999–2001 | 26:1 |

Not surprisingly, this caused nitrogen-fixing Cyanobacteria to proliferate, just as in lakes. Nitrogen-fixation of up to 2.25 milligrams per square metre per year was recorded by Brattburg (1986). Such values are even higher than in Lake 227!

Between 1968 and 1973, STPs in the area began P removal. As a result, P inputs from point sources to the archipelago had decreased by nearly 10-fold from over 700 tonnes per year in the late 1960s to about 75 tonnes per year by 1980 (Figure 13.5; Brattberg 1986). By 1980, only about 25% of the P inputs were from STPs discharging directly into the archipelago. Most of the P still reaching the archipelago was from the Lake Mälaren outflow. Since that time, there have been substantial year-to-year variations in P inputs, primarily because of differences in outflow from Lake Mälaren (Figure 13.5). A further decrease in P from STPs began in 1996, and removal of N from effluents began about the same time. Currently, only 15–20% of the phosphorus entering the archipelago is from STPs.

As phosphorus inputs decreased in the 1970s, so did the concentration of phosphorus in the archipelago. In the middle of the archipelago, concentrations of total P declined from 80–100 micrograms per litre in the late 1960s to less than 30 micrograms per litre by the mid-1970s. However, at

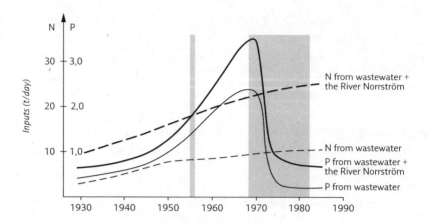

FIGURE 13.4. *Nitrogen and phosphorus inputs (t/day) from 1930 to 1985 to the Stockholm Archipelago from sewage treatment plants and from Lake Mälaren (through the River Norrström).* Figure from Brattberg (1986). Printed with permission from Vatten.

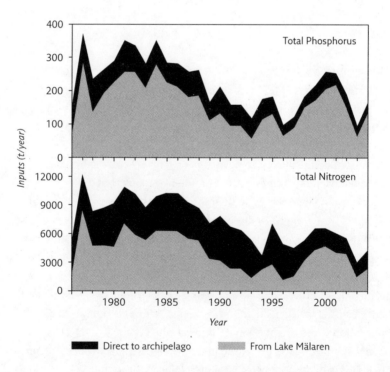

FIGURE. 13.5: *Phosphorus and nitrogen inputs to the Stockholm Archipelago from 1976 (well after P removal from STPs) through 2004. Inputs from STPs directly to the archipelago are in black and inputs from Lake Mälaren in grey.*

Figure is replotted from data in Lännergren and Eriksson (2005).

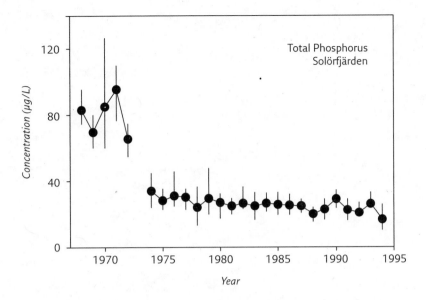

FIGURE 13.6: *Total P concentrations (µg/L) at Solörfjärden in the mid-Stockholm Archipelago between 1968 and 1994.* Data from Lännergren (1994).

that point, phosphorus reached a plateau, staying more or less constant through the 1980s and early 1990s (Figure 13.6).

The large blooms of Cyanobacteria declined and phytoplankton, measured as chlorophyll, decreased rapidly (Figure 13.7). Chlorophyll concentrations in the inner archipelago declined from over 30 micrograms per litre in 1969–1972 to about 16 micrograms per litre between 1972 and 1989. Overall, the results are as convincing a case for phosphorus removal as any whole-lake experiment; however, like phosphorus, chlorophyll also plateaued in the 1980s and 1990s, at values still well above pristine conditions. This hysteresis in recovery has caused much recent controversy among scientists and policy makers.

Some Baltic scientists have recommended that reducing point sources of nitrogen as well as phosphorus will be necessary in order to get further improvement in the archipelago. They point out that sediments have remained anoxic, recycling phosphorus efficiently back to the water column, for the same reasons that we discussed in Chapter 10. Others question how nitrogen removal could possibly be effective, because the ratio of total nitrogen to phosphorus has remained over 30:1, clearly in the phosphorus-limited range. Also, they believe that there is evidence that the current anoxia and phosphorus release from sediments are beginning

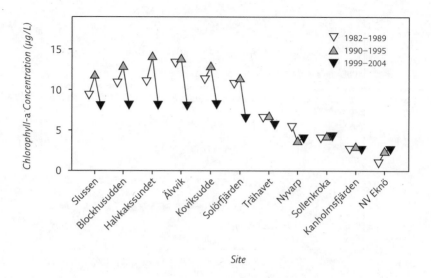

FIGURE 13.7: *Average chlorophyll* a *concentrations in the Stockholm Archipelago (left to right = inner to outer regions) during various time periods, indicating a decrease in chlorophyll, particularly in the inner regions, in the period 1999–2004 relative to previous time periods.* Data from Lännergren and Eriksson (2004).

to dissipate and that a time-lag similar to that observed following reduced external loading in lakes should be expected for the sediments to respond. However, the proponents of nitrogen control argue that the concentrations of dissolved inorganic nitrogen and phosphorus in surface waters in summer are both undetectable, so that the system could be nitrogen limited.

The nitrogen-limitation school of scientists persuaded Swedish regulators to reduce inputs of nitrogen from STPs to the archipelago since the mid-1990s. The result has been a large reduction in total nitrogen concentrations in the inner archipelago. However, there was also a decrease in P release from STPs at that time. Chlorophyll decreased slightly in the 1999–2004 period relative to earlier years, but it is impossible to tell whether reductions in N or P or both were responsible. Our guess is that it was the phosphorus. Phosphorus reductions had caused decreased algae earlier, when the system was nitrogen limited. Logic says that, because the system is now phosphorus limited, reducing phosphorus further would be even more effective than it was earlier.

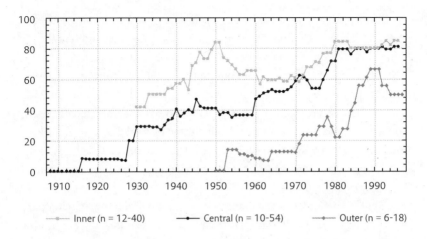

FIGURE 13.8: *Percentage of area covered by laminated sediments in the outer Stockholm Archipelago.* Figure from Jonsson (2003), and printed with permission.

It is believed by the nitrogen school that return of phosphorus from sediments is limiting the effectiveness of further reducing in phosphorus from external sources. However, we believe that there are signs that the release of phosphorus from sediments of the archipelago may finally be subsiding. There has been higher oxygen and lower phosphorus in deep waters in recent years, particularly in the inner archipelago.

Another possibility is that the removal of organic material rather than nitrogen at the STPs is responsible for the slight improvements noticed in oxygen and phosphorus. Organic matter consumes oxygen, promoting the release of dissolved inorganic phosphorus from sediments. If this is the major reason for declining phytoplankton, simply oxidizing ammonium and organic nitrogen to nitrate in the STPs before releasing it to the archipelago should have the same result as removing nitrogen, for far less cost. At present, data are insufficient to test this possibility.

Another source of controversy has been the appearance and persistence of laminated sediments in deep waters of the archipelago. The extent of these has increased over the years (Jonsson et al. 2003). Laminated sediments due to the absence of sediment mixing by benthic animals are one indication of the occurrence of bottom water with very low oxygen. They do not necessarily indicate permanent anoxia because even a few short low-oxygen events can kill the benthic fauna, drastically decreasing sediment mixing.

The area of laminated sediments has increased slowly. By the 1950s, the extent of laminated sediments was almost 80% of the inner, 40% of the middle, and 10% of the outer archipelago (Figure 13.8).

Proponents of N control point out that, even following the decrease in P loading in 1971 and the subsequent decline in phytoplankton, the laminated sediment area increased from 60% to about 80% in the inner archipelago by the mid-1980s. The increases in middle and outer regions slowed during the 1980s and stopped increasing in the 1990s (Jonsson et al. 2003). From about 1990 until 1996 or 1997, the extent of laminated sediments remained relatively constant (Figure 13.8).

There are now a few signs that recovery may be beginning. In outer parts of the archipelago, laminae disappeared at some sites after 1997, indicating that benthos had returned, mixing the sediments once more. In particular, Eckhéll et al. (2000) observed the return of the amphipod *Monoporeia* in response to improving oxygen conditions in recent years. At present, data are too scarce to assess how far the recovery might have proceeded, but clearly, the extent of laminated sediments had ceased increasing, and even begun to decrease, before the control of nitrogen inputs of the mid-1990s.

A long time delay is expected before a reduction in the return of phosphorus from sediments subjected to decades of enrichment, as we discussed earlier for the recovery of lakes (Chapter 10). The near-surface sediments remain rich in organic matter that supports high decomposition for many years after nutrients are controlled (Chapter 10; Søndergaard et al. 2002).

In summary, the Stockholm Archipelago initially recovered rapidly following decreased loading with phosphorus in the 1960s and 1970s, despite showing evidence of N limitation at the time. $N_2$-fixing Cyanobacteria also declined as the estuary became increasingly phosphorus limited. Although further improvement in water quality is desirable, we do not believe that controlling nitrogen will be beneficial. Rather, further reductions in phosphorus seem most likely to be effective. Effluents from the many non-point sources in the archipelago deserve particular attention. Also, the internal loading of phosphorus will probably occur only slowly, as we have seen is the case in many lakes. Decreasing inputs of nitrogen is unlikely to speed the recovery process. In contrast, it may favour the very N-fixers that are one of the most undesirable symptoms of eutrophication. As in the case of freshwaters, patience is required, because many of the symptoms of eutrophication are expected to persist for many years after external sources of phosphorus are decreased. Meanwhile, debates continue to rage, the result of the absence of clearly designed whole-ecosystem experiments.

# 14 | The Eutrophication Problem as Part of a Greater Environmental Crisis

*Solving eutrophication is the first step in solving a much greater problem faced by humanity*

EUTROPHICATION is but one manifestation of a much broader syndrome that is worrying ecologists all over the world: the overuse and over-pollution of planet Earth's resources. This book would not be complete without examining eutrophication in this broader context.

It is now over two centuries since humans first realized that they were having detrimental effects on our planet. The first widely known warning was sounded in 1798, when Thomas Robert Malthus published his "Essay on Population," the first documentation that human populations were beginning to undergo runaway population growth. Throughout the early 20th century, much concern was expressed about the rate of human population growth.

Already in the 19th century, there were signs that the human assault on Earth was much more than population growth. These were noticed by a few scientists of that time. George Perkins Marsh noted that as industrialization of the western world accelerated, so did detrimental effects on the environment, such as clearing land for agriculture, deforestation, canal building, water pollution, loss of natural species, and many other facets of nature. His 1864 book *Man and Nature* is widely regarded as the beginning of the environmental movement. About the same time, Angus Smith first documented the widespread acid rain caused by dispersal of airborne industrial emissions. By the end of the 19th century, Svante Arrhenius had deduced that increasing carbon dioxide from human sources was likely to warm the entire planet. He calculated that a doubling of carbon dioxide would cause Earth to warm by 5-6°C, a number remarkably similar to estimates published by the International Panel on Climate Change in 2007. However, Arrhenius could not realize how quickly rapidly industrialization would change the atmosphere. He estimated that a doubling of its atmospheric concentration would occur in 3000 years! We now know that a doubling of the carbon dioxide present in the atmosphere in Arrhenius's time will occur before 2050 unless we change our ways.

These early signs that humans were changing the planet's life support system were not widely known outside scientific circles. As a result, they were not noticed by most citizens until much later. More recently, advances in scientific instrumentation made it possible to measure the rate of change in several barometers of the Earth's condition. Charles Keeling's direct measurements of increasing carbon dioxide in the atmosphere began at Mauna Loa, Hawaii, in 1958. Modern chemical methods were developed for measuring increasing eutrophication and acid rain, and concentrations of other pollutants in the environment. Mid-20th century news media, such as television and computers, and a more educated press spread the word about insults to Earth very rapidly. Many believe that publication of the first pictures of planet Earth from space, taken by William Anders on the 1968 Apollo Mission, marked the beginning of an attitude change by human earthlings. It became obvious to most that Earth was small enough that the entire planet might be threatened by human actions. Later, satellite images showed the rapid destruction of the tropical rain forests that are known to produce oxygen and absorb $CO_2$, the "lungs of the planet" as some put it. Other recent human actions that were discovered to have global or broad regional consequences include the discovery of the stratospheric ozone hole in the 1970s and the realization that acid rain and eutrophication were happening over large areas of the planet. Rachel Carson's *Silent Spring*, published in 1962, documented the devastating effects of pesticides drifting over large areas of the planet. Even earlier, Charles Elton (1958) documented an even greater problem in his book *The Ecology of Invasions by Animals and Plants*, describing how introduced alien species were changing the Earth's ecosystems. Curiously, it did not get much popular attention at the time. Biodiversity is now known to be decreasing all over the planet, largely as a result of destruction of the habitats of animals and the encroachment of the alien invaders that Elton documented. Ecologists tell us that we are now in the midst of the worst mass extinction since massive meteors hit the Earth 65 million years ago.

The above were but a few of many studies and technological advances that made it obvious that human population growth alone is not the threat to our planet. The real threat to Earth is population growth coupled with increasing per-capita use of technology and energy, increasing per-capita discharge of pollutants, and per-capita damage to Earth's "ecological services," such as cleaning the air and the water. All are increasing exponentially, and it is the confounding effect of exponential growth and exponential consumption that has put our planet in crisis. The runaway combination has been termed *demotechnic growth*. (The runaway demotechnic cycle (*demos,* population; *techne,* technology) refers to a runaway cycle in which population and technology spur each other to greater and less ecologically sustainable heights). According to recent estimates, human consumption is already exceeding the carrying capacity of the planet by 23% (www.ecofoot.net/).

There have been several recent attempts to get this concept across to citizens, politicians, and business leaders. Probably the most successful has been the concept of an *ecological footprint,* a term coined in 1992 by William Rees of the University of British Columbia. An individual's ecological footprint is defined as the area of land that would be needed to provide the resources and absorb the pollution caused by the actions of the society or its individual members on a sustainable basis. Rees's book, *Our Ecological Footprint: Reducing Human Impact on the Earth*, authored with student Mathis Wackernagel in 1996, provides a detailed description of what calculating the footprint entails and the vast range in the requirements of different societies. For the 2006 population of Earth, 1.8 hectares of land were available to support each human. But average human consumption in that year required 2.2 hectares. Average North American consumption was over 7 hectares per citizen! Clearly, to accommodate projected growth in the Earth's population and the aspirations of Third World societies to western lifestyles, we will soon need the resources of several planets as rich as Earth to accommodate humans, but this is not possible. The only available solution is to reduce human demand or lose the ecological services of our mother planet.

There are many tough decisions to be made if human society is to survive on the one planet that we have at our disposal. These go well beyond the realm of science into the realm of politics, social science, ethics, and religion. However, science is needed to provide the basic understanding of the rates and the environmental price that we will pay for not changing our ways. For a more complete discussion of this runaway demotechnic cycle and its effect on the planet and human well being, see Vallentyne's (2006) book *Tragedy in Mouse Utopia*. Clearly, bringing our society back into harmony with what Earth can provide is the greatest of modern challenges. Several recent works have discussed in much detail what we must do to save Earth. DWS's favourite is Lester Brown's 2006 book *Plan B 2.0: Rescuing a Planet Under Stress and a Civilization in Trouble*.

In addition to the transformation of the media in the last 50 years, scientists began to emerge from their Ivory Towers and speak publicly about the Earth's impending problems. Before the mid-20th century, participation in social or political debates had been regarded by most scientists as unprofessional; this is one of the reasons why knowledge of the troubling state of the planet was so long in emerging. Today, such attitudes seem absurd to us, but many scientists still operate in the old-fashioned way.

### Solving Eutrophication: A Good First Step

EUTROPHICATION RELATES to many aspects of demotechnic growth via increased use of fertilizers, increased land clearing, increased use of cleaning products, increased culture of livestock, and increased building of summer homes on lakeshores. Perhaps, controlling eutrophication will give us the confidence and will that we need so badly to control other aspects of runaway demotechnic growth. It seems a good starting point, because water is one of most threatened of all of Earth's resources. Roughly one-third of Earth's humans are threatened by water scarcity or by water that is too polluted to drink or bathe in.

The solution to eutrophication seems simple enough: preventing excess nutrients from entering lakes; keeping the inflows, outflows, and wetlands in the catchments of lakes intact; and allowing the food chains and fish habitats of lakes to remain in their natural states. None of these seems very difficult to do; after all, we are in an age when "sustainability" is supposed to underpin all of our interactions with natural ecosystems. It seems reasonable to ask, *"With thousands of scientific studies of the problem, why do we not simply act to prevent or reverse the eutrophication problem?"* There are several answers. One is the fragmented way in which human societies govern lands and waters, where quick profit and local political advantage outweigh ecosystem protection. There is very little control of lakeshore development in much of North America. Every spring when people start thinking about lakes, billboards and newspapers advertise new massive lakeshore developments, where a summer home can be obtained that has all the luxuries and comforts of one's permanent home. Another reason for the lack of action is that the necessary scientific knowledge to make sound decisions is unknown to decision makers. As a result, human societies have often adopted simplistic, partial solutions where complex, holistic ones are required. In this way, eutrophication is also symptomatic of the wider demotechnic problem.

## Society's Control of Nutrient Sources is Very Selective

Point sources of nutrients are easy to control, as is now obvious to the reader. Society should expect waste from CAFOs to be subjected to treatment similar to that from cities of similar nutrient output. For the most part, this is still not done.

Control of non-point sources of nutrients is also necessary, but controlling non-point sources can be very difficult. Agricultural wastes and commercial fertilizers can be widely dispersed and tend to move toward water largely during large rainstorm or snowmelt events. Discharge from septic tanks moves underground, so that where nutrients end up is difficult to detect. As we have pointed out in Chapter 10, protecting wetlands and riparian areas helps to prevent transfer of agricultural wastes to waterbodies. There are other ways to dispose of wastes from isolated cabins than septic tanks, such as composting and combustion toilets. In short, society must adopt holistic approaches to controlling nutrient sources.

## Decision Making is Fragmented

OFTEN, the management decisions necessary to protect lakes and rivers need to include several levels of government. Decisions to approve lakeshore developments, expansions of cities, and new industries to draw water from rivers or aquifers are typically made at the municipal or county level. Little thought is given to other stresses on the lake or river. Developments are almost always justified by explaining that they are small, that they require very little water, or that they are essential for prosperity of a community. The sum of all of the little demands on a waterbody is often very large, but such "cumulative effects" are rarely considered. Often, municipal decision makers do not even consider scientific evidence, only whether the development will be good for local merchants, the local tax base, or local recreational opportunities. The effects of development on nearby or downstream freshwaters are not discussed. In one recent case, a small prairie city in Canada fought to keep huge hog farms out of the watershed that drained to its water supply but approved a slaughterhouse downstream on the same river! As we write, citizens of Devon, Alberta, have just lost a battle with regulators to prevent a large dairy farm from locating immediately upstream on the North Saskatchewan River, the water supply for the community, as well as for the million plus people who live in the Edmonton area. Ensuring that local interests do not override the health of an entire waterbody is the responsibility of state or provincial agencies, or in the case of large systems or international waters, of federal governments. Tragically, most often, provincial, state, and national regulators do not act until water-quality guidelines have been violated or fisheries decline, at which point developments are usually complete. In Chapter 10, we described briefly the low priority given by the U.S. Environmental Protection Agency (EPA) to nutrient management until very recently under the Clean Water Act. It is the responsibility of higher levels of government to ensure that strong, clear regulations are in place to guide lower levels of government and individual landowners so that their developments do not harm water resources. Such regulations should be strongly enforced and underscored by fines or other penalties.

When different departments and levels of government do take action, they usually end up in a jurisdictional mess. As we mentioned earlier, water quality in Canada is the responsibility of Environment Canada, but the health of fisheries is the responsibility of the Department of Fisheries and Oceans. The causes of the problem can lie within the jurisdiction of

Agriculture Canada or Natural Resources Canada. Health Canada or Indian and Northern Affairs is often responsible for the outcomes of management. For waters that lie within provincial boundaries, much of the responsibility for managing freshwaters has been delegated to the provinces. Here again, different aspects of water protection and water use fall to different departments within the government. In turn, the decisions that affect water quality are often left to municipal governments. The result is long delays, unsatisfactory compromises, and questionable environmental solutions. There is great reluctance to remove developments once they are in place, no matter how damaging they may be to the environment, so mitigation measures are usually unsatisfactory and almost always costly. Overall, the system is rather like closing the barn door after the mare has been bred! Once a lake becomes eutrophic, the scientists are called in to remedy what is by then a difficult and expensive problem to fix.

At present, there are some huge discrepancies in how laws apply to nutrient sources. For example, as Dr. William Lewis has pointed out, under the U.S. Clean Water Act, construction practices are now strongly regulated, and it is expected that silt and nutrients released by land disturbance during construction will be intercepted. In contrast, there are few rules regulating septic tanks or the fertilizing of lawns. Although the legal authority for doing so is implicit in the Clean Water Act, the necessary intrusive actions would be poorly accepted by a public not fully aware of the consequences of nutrients in water. The desire for green lawns still outweighs the public recognition of the causes of green waters.

There are a few signs that this is changing. Under Alberta's Water for Life Strategy, committees that represent whole watersheds have been appointed to produce "State of the Watershed" reports. It is still unclear whether the findings and recommendations of these committees will be well used to guide decision making, but it is a good first step. As we write in 2007, there is a proposal to divert water from the Red Deer River basin in Alberta to promote development of a gambling casino in the water-scarce Bow River basin, even though such inter-basin transfers are explicitly discouraged in the first paragraph of the Water for Life Strategy. Provincial officials explain that the Red Deer River and the Bow River are parts of the same South Saskatchewan River system. This explanation does not satisfy residents or city officials of the communities in the Red Deer Basin, who believe that they have the right to use and the responsibility to protect the waters of "their" river. It also defies logic: using such criteria, one could argue that diverting the Red Deer River into the Winnipeg River would be fine, because they both eventually merge in the Nelson River! It remains to be seen whether the Water for Life Strategy will significantly improve water conservation in the province. So far, a number of actions, such as the dairy barn described above, indicate that Water for Life is another toothless paper document.

There is also some progress in the United States. The EPA is requiring states to produce numeric standards for concentrations of phosphorus and nitrogen in U.S. waters. The standards must be approved by the EPA's judgement on values that will protect waters from impairment. This is a significant improvement from the early days of EPA, when individual states were left to make their own decisions on whether or not to regulate point sources of phosphorus (see Chapter 8).

### The Science of Watershed Management is not Communicated to Policy Makers and Citizens

WHAT SOCIETY CONSIDERS TO BE watershed "management" could possibly be more accurately described as an art. It is at best scientific "fine tuning;" when we use the term "manage" below, it is in this sense. "Management" seems to be a term rooted in the egocentric days when industrial humans were still ignorant and arrogant enough to believe that they could control the major ecosystems of the planet, including fresh-water. We should now know better, with over one-third of the planet's human residents faced with water that is unfit for drinking and sanitation. Older aboriginal people know better. In the words of Sal Martin, a Cree Elder "Water is the boss." However, while scientists can't be expected to magically solve all of the problems that have been ignored for years, what they know is still of considerable help to decision makers.

Much of the problem is one of failure to communicate the relevant scientific information on eutrophication to regulators and to the public. Key scientific works are still typically published in technical journals, where they are read by a small handful of experts who can locate them in the complex libraries of their Ivory Towers. Often, they are written in special-ized scientific jargon only understandable to limnologists. The failure to communicate is not entirely the fault of scientists. It is difficult to interest editors of newspapers and lay magazines to carry scientific stories, even those that are urgent and written in plain, understandable language. As those of us who try to communicate with decision makers know all too well, the efforts are often not welcomed by officials who have taken a stance that the science puts into question: there is a tendency to "shoot the messenger" rather than to hear the message. Scientists, even though they try to be objective and to base their conclusions on well-designed studies, are often dismissed as more unwanted environmentalist advocates. Many land developers, feedlot and golf course operators, city officials, and others who stand to profit or obtain tax revenues from the business-as-usual procedures that supply nutrients to lakes often try to discredit the science rather than take costly measures to protect freshwaters. Their tactics are easy to see through and common to many types of industry, as pointed out by the Union of Concerned Scientists (2007). It takes a scientist with a thick skin to cope with all of these factors, which are unrelated to the reasons why most scientists choose science as a career. As a result, a more typical scenario has been for a scientist to publish his work in an academic journal, which is only read by other scientists with similar specialties. If it amuses or annoys enough of them, the article will be quoted by others. These "citations" are considered as evidence that a scientist should be promoted, not whether the work contributes to solution of a problem.

Fortunately, once again positive changes are occurring. In both the United States and Canada, watershed associations are becoming very common. Interested citizens, local government representatives, and even higher officials can usually be involved. The need to manage whole watersheds is easily grasped by most citizens, and cumulative problems such as nutrient loading and the effects of upstream developments on downstream users are recognized. In recent years, the Rosenberg Water Forums have been developed to get regulators, scientists, and politicians all in the same room at the same time and have candid discussions of what might be done to solve water problems quickly and efficiently. Hopefully, such methods will soon be used for other aspects of the demotechnic problem, because rapid action is now human society's only recourse.

## Holistic Watershed Management

It is obvious from the above that, to protect freshwater supplies and fisheries in the coming century, a much more comprehensive approach is needed. Simply controlling nutrients from detergents and a few point sources will be insufficient to prevent eutrophication in countries with increasing populations, agriculture, and technology, especially when a warming climate is threatening to reduce water quantity and aggravate problems of water quality. Although it is important to continue to improve nutrient removal, humans must also protect better the catchments that supply their water. Limits must be placed on the amounts of nutrients and other chemicals imported to these catchments, and the location and timing of their use. Wetlands and riparian zones must be protected, restored, and maintained to minimize the chance of nutrients getting into surface and ground waters. These measures will require comprehensive planning of human activities in the catchments that supply their water. Unfortunately, catchments typically cross into several political jurisdictions, because the flow of water does not respect interdepartmental, interstate, interprovincial, or international boundaries. Early in the 21st century, decisions that affect the water quality of all downstream waters are typically still left largely in the hands of individual municipalities. Effective water management at catchment scales should involve individual municipalities, industries, and land-use changes strategically considered with the benefit of all residents of the catchment, including natural biota, in mind. It is clear from Chapters 9 and 10 that maintaining the health of complete aquatic ecosystems is essential to maintaining water quality. In short, fisheries should be regulated to protect water quality as well as to protect fisheries. Overall, good water quality requires improvements in the management of many activities in the catchments of waterbodies, and of fisheries.

### The Belief in "Magic Bullets"

POLICY MAKERS ARE OFTEN CONFIDENT that we can proceed with most any kind of development, fixing the problems as we go. As we have seen, this is almost always more expensive and time consuming than simple prevention, yet it is still the usual way that developments proceed.

Equally absurd are some of the proposed methods for masking, postponing, and bypassing the symptoms of eutrophication. At least two American companies make blue dyes that can be applied to lakes or ponds to make their green eutrophic waters appear to be pristine! Fortunately, such cosmetic measures are not widely used for public waters. There are bubblers that are supposed to substitute for the natural mixing processes that maintain oxygen and prevent nutrient return from sediments. Chemicals to reduce phosphorus return can be harrowed into sediments, or nutrient-rich sediments can be physically removed by dredging. Floating weed harvesters cut unsightly weedbeds, but the weeds return the following year. Fortunately, after being widely used for about 20 years,

weed harvesting is now used mostly for maintaining navigation canals.
In Canada, cutting aqautic weeds without a permit can be regarded as
destruction of fish habitat, a violation of the federal Fisheries Act. All of
these techniques can be bought from or applied by specialized industries.
However, none of them are very effective, and all are expensive; ultimately,
the sources of water and nutrients must be controlled before lakes can
recover, as we described for Lake Tuusala in Chapter 12.

### Water Transport of Wastes: Valuable in the Past, but is it Still the Option of Choice in the 21st Century?

THE FIRST LINK connecting the immediate to the ultimate causes of
man-made eutrophication is the recognition of an intermediate connection—the role of the water transport system of waste disposal. When
humans were dispersed as small bands of hunter-gatherers, there was no
real need for an engineered wastewater system. Nature took care of the
nutrients and pathogens. Indeed, many of the earliest agricultural societies recognized human wastes as a valuable commodity, which could
be deliberately applied to crops to enhance their growth. As populations
began to aggregate and increase, simple dry pit toilets became popular.
The excrement of livestock largely replaced the need for recycling human
excrement to fertilize crops. However, as populations increased and cities
grew, the amount of waste generated by the "pathological togetherness" of
humans and their domesticated animals in small areas became unaccept-
able. Increased waterborne disease and horrible odours were the result.
It was in response to the growing waste problem in cities that engineered
waterborne waste disposal systems were first introduced on a major scale
in the latter part of the 19th century. It is not surprising that, although
these engineering systems were effective at removing waste from cities and
reducing waterborne diseases, they permitted the eutrophication problem
to spread. As long as cities were few and small, this system sufficed. Early
symptoms of eutrophication were rather mild, and the distances between
human communities along rivers were great enough to allow the biotic
communities to render the water clean enough to drink again. Overall, the
purpose of the water transport system of waste disposal was to improve
health. To those suffering from cholera, diarrhea, and other waterborne
disease, improved health seemed well worth the cost of the water-based
system.

In order to appreciate the extent to which improved health was needed in the 19th century, it is necessary to understand conditions existing then, particularly in densely populated areas. Infant mortality rates in urban centres of Europe and North America were commonly double those in adjacent rural regions, and both were 10–20 times greater than those existing today. In the larger cities, one out of every four children died before the age of one year. Of the remaining three, another would depart before the age of five. Of those who survived with a debilitating sickness or physical injury, it might easily be said by others that they would have been better off dead.

Two quotations from Henry Jephson's book, *The Sanitary Evolution of London*, serve as a reminder of what life was like in London, the thriving capital of the British Empire, in the middle of the 19th century.[1] Jephson, a responsible member of London City Council, quotes Mr. J. Phillips, who appeared as a witness before the Metropolitan Sewers Commission in 1847:

> *In pursuance of my duties, from time to time, I have visited very many places where filth was lying scattered about the rooms, vaults, cellars, areas, and yards, so thick, and so deep, that it was hardly possible to move for it. I have also seen in such places human beings living and sleeping in sunk rooms with filth from overflowing cesspools exuding through and running down the walls and over the floors.... The effects of the stench, effluvia, and poisonous gases constantly evolving from these foul accumulations were apparent in the haggard, wan, and swarthy countenances, and enfeebled limbs, of the poor creatures whom I found residing over and amongst these dens of pollution and wretchedness.*

In discussing the dismal state of water supplies (all provided by private companies at the time), Jephson quoted extensively from unpublished reports of various Medical Officers of Health:

> *From Shoreditch (1860), the Medical Officer of Health wrote: "I have hardly ever exposed a sample of town spring water to the heat of a summer day for some hours without observing it to become putrid." In St. Giles (1858–59), "the water of the wells was not deemed good enough (on analysis) for watering the roads." In St. Marylebone, "44 public wells supplied water which was for the most part offensive to taste and smell." In Kensington (1860) "all the well waters of the parish were foul." In Rotherhithe (1857), "The water from the tidal well smelt as if it*

*had recently been dipped from a sewer." The Medical Officer of Health for Lambeth declared (1856) that "the shallow well waters of London combined the worst features...they represent the drainage of a great manure bed."*

*The people were driven to the use of the water from these wells owing to the deficient and intermittent supply of water by the various Water Companies...water supplied for less than an hour a day by one single standpipe in a court containing hundreds of people...water supplied every second and third day, and none on Sundays, the day of all others on which it was most wanted; and the house-owners had provided no cisterns or reservoirs of proper capacity, and the Vestries had not compelled the house-owners to do so.*

*In some parishes, hundreds of houses had no supply at all. In some houses which had a supply, the tenants were deliberately deprived thereof by the Water Companies, because the house-owner had not paid the water-rate.'"*

After reading such descriptions, are there any humans so devoid of compassion for their fellows that they could resist improved measures for more sanitary living conditions?[2]

As outlined in Chapter 2, the introduction of the water transport system of waste disposal resulted in four undesirable consequences: (*i*) increased water pollution from physiological wastes because of the by-passage of soil; (*ii*) increased production of fertilizers to replace nutrients removed from the soil as crops, passed through the intestines and kidneys of man, and delivered to sediments and the sea; (*iii*) increased cost of treating wastes composed of more than 99.99% water so that waters downstream of waste treatment plants would again be usable; and (*iv*) man-made eutrophication.

With all of these features now recognizable, it seems worth asking again whether the systems used for nutrient management in the past still make sense today. As we have shown, making and distributing phosphate and nitrogen fertilizers has become big business. Lands have been cleared and livestock cultured not only to supply local populations but also to foreign markets in areas where water or good agricultural land are too scarce to support local populations. However, the full costs of this wasteful system are seldom summed. The costs of fertilizer, treating water, battling eutrophication, and the medical costs of waterborne disease are kept separate from the profits made by industry and agriculture and paid for through taxes. If society as a whole does not profit from catchment development, the system becomes rather senseless, analogous to living only by paying attention to your income while ignoring your expenditures. There has been a recent trend toward "full cost accounting" in order to more realisticallly balance the benefits and costs of human activity, but it is still rarely applied.

The introduction of waterborne waste disposal to protect public health has permitted another consequence to come about, more costly than all four of the previously mentioned consequences combined: *urbanization,* a problem that has yet to reach its fullest and most revengeful expression. By removing a major health barrier to increased population density, all of the economic and behavioural forces that cause humans to cluster together are unleashed. In this sense, a fair share of every unwanted attribute of urbanization—slums, smog, overcrowding, noise, traffic congestion, murder, riots, alienation, and lead poisoning in children who nibble at old wall paint—can be attributed collectively to installation of improved measures for urban sanitation, simply by reducing the probability of disease that would otherwise be present.

## Some Practical Solutions to Eutrophication

### Protecting Nature's Plumbing

One can view lakes, rivers, and aquifers as the plumbing of the landscape. Catchments drain into these waters, making them more or less analogous to sewers in houses. Yet these "sewers" of the landscape are where people draw water for drinking and sanitation and where they seek fish and other aquatic foods. This view makes it obvious that human societies should strive by all means possible to protect these waters, both maintaining their flows (or in the case of lakes, levels) and keeping nutrients and other pollutants out.

There is some "good news" to report here. What is needed is not some mysterious new approach that cannot be implemented until billions of dollars are spent on new scientific studies. Quite the opposite is the case. Most of what people need to know to protect their watercourses is readily available in Ivory Tower libraries in the form of literally thousands of research papers. The problem is that those who make the decisions that affect the fate of the world's waters are unaware of the science, or if they do know it, they are unwilling or unable to implement it.

The measures that need to be taken are quite simple. The first is to leave wetlands and riparian (streambank and lakeshore) areas in their natural states. In populous parts of Canada and the United States, 70% and more of wetlands have been removed. In the United States, the value of wetlands was made visible about two decades ago, through implementation of section 404 of the Clean Water Act. Some officials and developers still argue that protection of wetlands is a waste of money and resources, but no current politician would dare openly espouse the elimination of the U.S. wetland protection system, although there are still attempts to weaken it by changing the definition of the term "wetland." In Canada, there is still no wetland protection. In most cities, developers simply fill them in or drain them to build housing developments or roads. In rural areas, farmers fill or drain them to allow more land to be planted or allow large equipment to be operated more easily. Riparian areas are treated similarly. There is still not widespread public recognition of the fact that wetlands and riparian zones filter out nutrients, silt, pathogens, and many toxic contaminants that would otherwise reach our lakes and rivers. Wetlands also recharge groundwater aquifers and act in a fashion analogous to capacitors in an electrical system, holding water on the landscape and releasing it slowly over time. In this respect, they are valuable for both flood control and nutrient removal. In short, protection of wetlands and riparian areas is badly needed. Incentives to landowners to protect these critical features of the water landscape would help.

More effort should be devoted to leaving natural drainage patterns intact. Road construction and paved areas disrupt natural flow patterns. Often, farmers use ditches and drainage tiles to get water off their fields faster, not realizing that widespread channelization and rapid drainage can promote flooding in downstream areas. In some progressive areas, roads, parking lots, and other paved areas are being designed to allow water to seep through into the soil below, rather than running off into lakes and rivers. Road crossings must be designed so as not to disrupt the flow of water, or the passage of fish.

Simple culverts or short bridges do not do a good job of protecting wetlands from roads, because the mass of wetland vegetation and peat restricts the flow of water horizontally. As a result, when driving along a road that crosses a wooded wetland, you will notice on one side the water level is higher than the other. On the high-water side, dead trees usually protrude from the water, a result of flooding. On the other side of the road, where water levels are lower than normal, tree growth accelerates.

Low tillage agriculture has helped to protect watersheds from water erosion as well as wind erosion, but some farmers still do not use it. Low tillage saves fossil fuels and costs, as well as protecting waters from inputs of silt, nutrients, and pathogens. However, in some cases, increased use of herbicides is used to compensate for decreased tillage. Uses of fertilizers, herbicides, and pesticides need to be well planned, with amounts and times of application carefully considered, to protect watercourses. Fortunately, it is usually the farmers themselves who benefit most from such conservation strategies, because the waters that they drink are the first to be polluted by bad agricultural practices.

These are some of the general measures that can keep natural flows intact, as well as minimize the amounts of nutrients, soil, herbicides, pesticides, and other contaminants that get into our watercourses.

## Dams and Diversions

In addition, societies need to re-think the use of dams and diversions as solutions to human water needs. It sounds wonderful to hold back water behind a dam to use it as needed, but this changes the morphology of the river below the dam. The floods that shape and re-shape the channels of most rivers are reduced. Also, reservoirs at northern latitudes are generally emptied during winter, re-filling in spring and summer when there is the greatest need for water in the river downstream of the dam for both biota and human activities. Reservoirs cause more of the water to evaporate because of their large surface areas. Diversions reduce flows in some rivers and increase them in others. Numerous studies have shown that dams and diversions are highly detrimental to aquatic ecosystems, causing siltation, changes in channel morphology, pollution of remaining fisheries with mercury, destruction of spawning areas, and displacement of aboriginal people who have lived for millennia on the undammed river. In the words of Daniel P. Beard, retired Commissioner of the U.S. Bureau of Reclamation, in the September 1996 issue of World Rivers Review:

> We spent billions of dollars over several decades to build scores of large dams. We have received benefits from those projects, and they have contributed to local and regional development efforts. But there is another legacy to our dam building.
>
> We have drained wetlands and destroyed thousands of hectares of biologically rich habitat supporting all manner of lifeforms. Our rivers

and streams are polluted, and their reduced flow from damming and diversions limits their ability to flush themselves of pollutants. Salts accumulating from irrigation have destroyed farmlands, and polluted agricultural runoff has become a critical problem in many regions. Once-productive fisheries are now just a memory in many places. And we are faced with the decline of numerous other freshwater-dependent endangered species.

As we wrestle with these problems, we are also learning how costly it will be to clean up and correct them. Even by the most conservative estimates, the costs will be substantial. We will ultimately spend tens of billions of dollars to address the legacy of our dam-building era.

This is the most important lesson we have learned from our water-development experience: we reaped great benefits, but there were very great costs. For some, the jury is still out whether the benefits outweigh the costs. But for many, the answer is simple: we have paid too dearly for "cheap" power and water.

These substantial costs are always overlooked by dam promoters. Like high-pressure salesmen, dam boosters paint an ideal world: Cheap power, cheap water, increased agricultural production, economic development and an end to hunger! The reality is somewhat less rosy.

Building a dam is the same as constructing a nuclear power plant: you get immediate benefits, but also long-term costs of a very great magnitude. In addition to the startling social costs, a dam can leave a permanent legacy of environmental destruction that will take generations to correct. Neither of these costs are adequately figured into the cost of building a dam.

A basic premise for the U.S. water development program was that those who benefited from dam projects should repay the costs of building the project—although not necessarily the costs of repairing the damage, which was mostly just ignored. But the reality was that powerful beneficiaries of large water projects used their political influence to transfer their costs to the average taxpayer. Farmers, cities, and power users who were supposed to pay all the costs ended up paying only a fraction of what they were obligated to pay. For example, the Bureau of Reclamation spent $22 billion to construct 133 water projects in the western United States. Approximately $17 billion was assigned to be repaid by users.

The vast majority of the benefits of these projects was received by irrigated agriculture. But using political pressure, farmers got their repayment obligation reduced to $7 billion. After further adjustments and special relief, that amount was still further reduced by nearly 50 percent, to $3.4 billion. Thus, farmers who were the primary beneficiaries of these projects repaid only about 16 percent of the costs at zero percent interest, over a 50-year repayment period.

We also found those who promoted dam projects were not honest about costs and benefits. Water projects were never built on time or under budget. In our experience, the actual total costs of completed projects exceeded the original estimated costs, including inflation, by as much as 300 percent. This does not include the costs we are now facing for rehabilitating ecosystems damaged by these projects.

In the US, water resource policies were originally conceived and implemented to meet the needs of agriculture and mining. That was an acceptable approach as long as there were ample water supplies for cities, plentiful government funds, and limited influence from "civil society." All that has now changed.

Water supplies are no longer plentiful, largely due to increased population and greater demand for new uses, including the original user—the environment. Public funds for dam projects are no longer plentiful. A whole host of environmental considerations are now taken into account before water projects can be built. An active public participation process designed to give citizens access to key information and require a fair and open discussion of the impacts of water projects is another factor impacting water choices. Finally, public support for financial subsidies to a small number of farmers or landowners, which had been the reason many of our projects were originally built, has disappeared.

These changes represent a significant shift in public opinion surrounding water issues. The American public now values low-cost, environmentally sensitive solutions arrived at through an open process where alternatives are fairly debated and information made available to everyone. This change in public opinion came about because of the important lessons we learned from nearly 100 years of building and operating dams.

After more than 50 years of unbridled dam-building, we've finally learned that there are many alternatives to solving water resource problems. These alternatives are often less costly to implement and have fewer environmental costs. For example, we now recognize the benefits of serious demand-side management and conservation; of using water pricing and metering to help allocate water fairly; of using

*comprehensive planning methods, and of open, inclusive decision-making*
*in stimulating new ideas and avoiding costly mistakes.*

*We are also beginning to repair some of the damage from our legacy*
*of river development projects. The two largest federal agencies respon-*
*sible for water projects are now taking action designed to make their*
*facilities more environmentally acceptable. The Bureau of Reclamation*
*is involved in tearing down a dam on the Elwha River in Washington*
*to restore a salmon fishery, and learning to operate other dams in ways*
*to help restore some of the damage to downstream ecosystems. The*
*Army Corps of Engineers is removing concrete channels from Florida's*
*Kissimmee River and restoring the original meanders, to improve the*
*river's ability to reduce floods. The Corps is now the leading Federal*
*agency for protecting wetlands—a major change from even just a few*
*years ago.*

*These new ways of thinking took nearly as much time and effort to*
*devise as it took to destroy our rivers in the first place. So when the*
*world's dam-builders turn to outside critics of their projects and cry foul,*
*one might ask, "What does this person stand to gain from this project?*
*And what does everyone else stand to lose?" For those of us now being*
*saddled with the costs of years of unquestioned dam-building, it would be*
*unconscionable to remain silent.*[3]

Fortunately, the era of big dam building appears to be over in the United
States, but Canadians have still not learned the bitter lessons outlined by
Beard. Canada has already impounded areas that together equal the size of
Lake Ontario. More diversions are planned, with little effective opposition,
despite the impacts on water resources and First Nations.

Finally, the potential effects of changes to the fisheries for promoting
eutrophication are enormous (Chapter 9), but most freshwater fisheries are
over-exploited. Freshwater species, including fishes, have suffered more loss
of biodiversity than any other groups on the planet. In first-world countries
where food is abundant, this is a sad commentary on human greed.

### A Water-Conserving Society

In addition to protecting the landscape's plumbing and aquatic communities, there is much that we could do as a society to decrease the amount of
water used by humans, and the amount of water that is fouled with nutrients and other contaminants. For example:

1.  Human societies could choose crops that require less water to grow. For example, canola (rapeseed) and small grains generally require less water than sugar beets or corn (maize).

2.  Society could demand state of the art technology for irrigation or reduce the amount of irrigation used for growing crops. Although most of North America has seen some improvement in the efficiency of delivery of irrigation water, some areas still simply flood fields, conduct water in leaky open ditches, and irrigate during hot, windy, daylight hours.

3.  Livestock culture could also use less water. The water use by the most efficient livestock operations is about 20% of the average, yet few incentives or regulations require farmers to adopt best practices in their operations. A few large livestock operations compost their manure, in some cases using the resulting methane to heat their farms. The manure is dried and sterilized during the process and, therefore, is much more amenable to transporting it away from the relatively small intensive livestock sites. In contrast, manure is costly to transport, because it is largely water. An alternative is to decrease the reliance on livestock for food, because of their high requirement for water, but not all humans wish to become total vegetarians.

4.  Society could require intensive soil testing as a prerequisite for application of manure or other fertilizer. Often, the amount of manure allowed is indexed to its nitrogen content. Most of the nitrogen applied is removed when crops are harvested. However, manure contains N:P at a weight ratio of about 4:1 or less, whereas plants require it at about a ratio of 15:1. As a result, excess phosphorus is left on the land each time manure is applied. Eventually, soils become so saturated that they release excess phosphorus to downstream waters for decades.

5.  Society could choose industries that require little water or that, at least, would return most of it to watercourses relatively free of pollutants. At present, this is seldom considered when industrial proposals are evaluated.

6.  A host of personal measures can be taken to use less water. From 300 to 450 litres per person per day are treated to human consumption standard in Canada and the United States. Only 1–2% of this is actually consumed...the rest goes down our sinks

and shower drains, flushes toilets, etc. Why not capture the water from sinks and showers and use it for flushing toilets, watering gardens, or hosing off the driveway? Another possibility is to have only a small amount of water treated to consumption standard, and the rest treated to a lower standard for other needs.

Measures for cutting water consumption are widely known but not widely implemented. For example, low-flow showerheads and low-flush toilets are always mentioned by green organizations. Not flushing toilets after every pee and not letting the water run while brushing one's teeth are not widely used. There are more complex solutions that individuals can use to lower their demands. A cistern can be used to collect water from shower drains and sinks, and used for watering lawns or gardens or flushing toilets. Elimination of water-demanding bluegrass lawns and watering when temperatures are cool, or using rain barrels or cisterns to collect rain water for gardening, are among the simple methods that people could employ. Collecting rainwater in urban areas has another side-benefit. It reduces the amount of water that floods into storm drains during thunderstorms, helping to prevent nutrients, pathogens, and toxic chemicals from being washed into watercourses. To ensure that their children benefit from healthy watercourses, human societies need to begin taking measures to ensure that human populations can be sustained by their environments, including water supplies.

*Is it Time to Change our Strategy for Waste Transport?*
It is time that human societies ceased using water to carry human wastes. Sewage and agricultural wastes now pose one of the greatest problems that must be dealt with in protection and treatment of water. It is not just a question of nutrients and pathogens. In addition, modern societies discharge dozens of pharmaceutical products, antibiotics, hormones, and personal care chemicals into sewage. Some of these are not efficiently removed by contemporary sewage treatment, so that they end up in water supplies downstream. The quantities of each are small, but the cumulative effects are unknown. For example, could they be in part responsible for the increasing antibiotic resistance seen in humans and animals? A recent study by Dr. Karen Kidd and her colleagues at ELA added small concentrations of the hormone in human birth control pills to a small lake containing

fathead minnows (*Pimephales promelas*). Most of the male fish were femi-nized, and reproduction all but ceased. Within 2 years, the population had almost disappeared. The fish population recovered once the hormone addi-tions ceased.

There are alternatives to using water for waste disposal that are reli-able, such as composting or combustion toilets. The resulting residue can be disposed of carefully, so that the amounts of nutrients and other chemi-cals that reach watercourses are minimized. In some parts of Sweden and Japan, systems of separating grey water, urine, and fecal matter for sepa-rate treatment have been used for decades.

Stringent water conservation and protection measures will take us far, but there must eventually be restrictions on human population and activi-ties imposed by freshwater, if freshwaters are to be healthy ecosystems that support natural ecological communities. This problem is approaching rapidly in the western world, and it has already arrived in the Third World. The time to begin considering the social changes necessary to protect freshwaters is NOW.

If human societies fail to heed the warnings provided by the many scientists who have filled library shelves with studies of the eutrophica-tion problem, the results are certain: *Our heads will be anointed with oil and algae when we go to the beach for a swim. Our cups will run over because no one will want to drink what is in them. And goodness and mercy will follow the Lord, but not us—for we shall dwell in the house of our wastes forever.*

# Appendix

On June 8, 2007, I exchanged several e-mail messages with Jack to discuss final details of this book. I was able to tell him that *The Algal Bowl* had just been officially accepted by the University of Alberta Press, which delighted him.

Six days later, I heard from Jack's son Peter that he was in the final stages of intestinal cancer and not expected to live more than a few days. Jack had known for several months but never mentioned it to me. Jack died peacefully on June 16, with his family around him. How like him to work until his final hours, pursuing his life-long mission of getting people to care for our planet! Jack's body was cremated and his ashes scattered over Lake Superior, which he loved dearly.

When I first met Jack, I was a young scientist who was having second thoughts about an academic career. The Ivory Towers seemed pretty tame to one in his mid-20s and seemed to involve a lot of posturing about complete nonsense. I needed some adventure in my life, as well as to feel that I was doing something of importance. Jack provided both, recruiting me to the Fisheries Research Board of Canada, and putting me in charge of starting the Experimental Lakes Area (ELA) project, which in the early years was as much an outdoor adventure as it was science. I sometimes wonder where I would be if Jack, and the ELA, had not come along when they did.

Thanks to Vallentyne, the early years at Freshwater Institute were truly the golden era of limnology in Canada, and perhaps in all the world, and I am grateful to have been a part of it. He put together a group so strong that even 30 years of suppression by bungling federal bureaucrats has not been able to throttle it. This year we will be celebrating the 40th anniversary of the ELA project.

Vallentyne was truly a great Canadian, who cared deeply about the welfare of his country and of planet Earth. His friends, colleagues, and disciples miss him greatly. —DWS

## John R. Vallentyne

*by D.W. Schindler (with contributions from G.J. Brunskill and Mohi Munawar)*

DR. JACK VALLENTYNE died peacefully in his sleep at his home in Hamilton, Ontario, Canada, on Saturday, June 16, 2007. After cremation, his ashes were scattered on Lake Superior, which he loved and cared for immensely. He is survived by his wife, Ann; children, Peter, Stephen, Jane, Anne-Marie and Geoffrey; and five grandchildren.

John R. Vallentyne was born in Toronto, Ontario, on July 31, 1926. Following three years of military service from 1943–1945, he finished his undergraduate degree at Queen's University, Kingston, Ontario, in 1949 and then went on to Yale University, New Haven, Connecticut, where he did his Ph.D. with G. Evelyn Hutchinson, finishing in 1952.

### The University Phase

Vallentyne's career as a limnologist and biogeochemist began in academia. He was a lecturer, then Assistant Professor at Queen's University from 1952–1958. He then moved to Cornell University, Ithaca, New York, as Associate, then Full Professor of Zoology, remaining there until 1966. According to his former students, during his tenure at Cornell, Vallentyne had catholic interests, including cosmochemistry, molecular evolution, pheromones, chemical ecology, paleolimnology and the history of science. During this period, he also spent a year at the Tonolli Laboratory in Italy as a Guggenheim Fellow.

*The Freshwater Institute Phase*

Following attendance at a Gordon Conference in 1966, where using whole lakes as test tubes was a topic of discussion, Vallentyne left Cornell to join W.E. Johnson in forming a new, as yet unformed, institute in Winnipeg, Manitoba, where whole-lake experiments with eutrophication were to become a major part of the Fisheries Research Board of Canada's freshwater program. Johnson was the Freshwater Institute's first director, and Vallentyne headed the new eutrophication section.

It was at this point, in January of 1967, that I (DWS) first met Jack. He invited me, then a young assistant professor at Trent University, to visit the new institute housed in a small building rented from the University of Manitoba. As we toured the empty building, Jack told me how he planned to fill it with the world's best limnological minds. I was cynical; to say this about Winterpeg, noted for its ferocious winters, and at a small university. You must be insane! I returned to Trent. A year later, Vallentyne phoned to persuade me to visit again. I was flabbergasted to learn that he had fulfilled his boast, recruiting Richard Vollenweider, Kazimierz Patalas, Arnold Nauwerck, F.A.J. Armstrong, and Domenico Povoledo from Europe and Mitsuru Sakamoto from Japan, to name the experienced scientists. There was also a large group of us in our mid-20s to early 30s, including Gregg Brunskill, Andy Hamilton, Ole Saether, Bob Hamilton, Stefan Holmgren, John Stockner, and myself. Soon to follow were Ev Fee, Bob Hecky, Bob Newbury, and Buster Welch. It was a group of tremendous intellect and energy, and the Freshwater Institute was a very exciting place to be.

The Experimental Lakes Area (ELA) was to be one of the new institute's foci, and I was rather speechless when Jack asked me, one of the youngest recruits, to lead it. However, the novel way in which Jack ran his section made it possible. In the early years, we had frequent meetings and numerous hot debates on new programs and new experiments. We were encouraged to be merciless and open in our critiques of the science of other members of the section. The philosophy, as Jack once put it, was "If you think your friends' critiques of your science are harsh, wait until your enemies get hold of it." Decisions were made by consensus.

In addition to the ELA, Jack established a Fisheries Research Board detachment at the Canada Centre for Inland Waters (CCIW) in Burlington, Ontario, to complement the physical and chemical programs of that institute. Vollenweider was appointed to head that unit. Others were assigned to the Regional Limnology group, to develop a broad knowledge of Canadian lakes.

Vallentyne was also the primary organizer of the first meeting of the American Society of Limnology and Oceanography in Canada (Winnipeg, 1971) and the first meeting of International Limnology Society (SIL) in Canada (Winnipeg 1974). He served as vice-president of the American Society of Limnology and Oceanography (ASLO) from 1964–1965 and president of SIL from 1974–1980.

Vallentyne became well known for his leadership in the debate over control of the eutrophication problem in the St. Lawrence Great Lakes. He used data from a large unpublished report by Richard Vollenweider and from early experiments at the ELA to clearly and successfully rebut a Madison Avenue-style campaign by detergent manufacturers and convince the International Joint Commission (IJC) to recommend regulation of phosphorus inputs to the Great Lakes.

### The CCIW Phase

Unfortunately, in the early 1970s the Fisheries Research Board was disbanded and its activities moved first to Environment Canada, then later to the Department of Fisheries and Oceans. Jack announced that he would be stepping down as section leader, because there would be little room for leadership within the civil service. He remained a Senior Scientist in the department until his retirement but moved his offices to the CCIW in Burlington in 1977, a more convenient location for his next activity, to co-chair the Great Lakes Science Advisory Board of the IJC. He influenced the IJC to change from a water-quality focus to an "ecosystem approach" that would include the watershed and airshed of the Great Lakes. During this period, he planned the public outreach for which he would become renowned in later life.

### The Johnny Biosphere Phase

On July 20, 1980, Vallentyne launched the Johnny Biosphere project. From personal conversations, I knew that he was becoming discouraged with the rate at which adults were solving Earth's environmental problems. He moved his hope to new generations, whom he hoped might be influenced by the introduction of environmental sensibilities early in their lives. Jack often appeared in public wearing a safari suit and carrying a globe on his back, shocking some of his more staid scientific colleagues. However, Johnny Biosphere delighted children worldwide, visiting up to 100 schools and 20,000 children a year where his basic message was "be kind to the

Earth and it will be kind to you." Johnny's website has gotten many thousands of visits, and it remains active for now.

*The Later Years*

In 2001, Vallentyne was awarded the ASLO's Lifetime Achievement Award for his tireless work in limnology and public awareness.

Active until his final months, in 2006 Vallentyne published *Tragedy in Mouse Utopia: An Ecological Commentary on Human Utopia*. He passed away shortly after our revised edition of *The Algal Bowl* had been accepted for publication by the University of Alberta Press.

The University of Alberta Press would like to thank the International Society of Liminology (SIL) for permission to reprint the obituary for John R. Vallentyne, originally published in *SILnews*, Volume 51, December 2007.

# Notes

## 1 | The Algal Bowl

1.  A white disc 8 inches (20 centimetres) in diameter used as a measure of water transparency, it originated with Commander Cialdi, head of the Papal Navy in 1865. On board the SS *L'Immacolata Concezione* (SS *Immaculate Conception*), he and Professor P.A. Secchi conducted a series of tests that led to the development and standardization of the "Secchi" disc. A modern Secchi disc is usually painted with alternate white and black quadrants, because the white/black transition lines make it easier to exactly tell where the disc disappears.

## 2 | Lakes and Humans

1.  Lake Erie was first seen by Europeans in 1669 when Louis Jolliet, guided by a friendly Iroquois, entered the lake from the west via Lake St. Clair and the Detroit River. The prime reason for the late discovery was the presence of the warfaring Iroquois. They controlled the portages between Lake Ontario and Lake Erie.

2.  When William Francis Butler passed through the St. Lawrence Great Lakes in 1870 on his way to help quell the Red River Rebellion in western Canada, he remarked, "But this glorious river system, through its many lakes and various names, is ever the same crystal current, flowing pure from the fountainhead of Lake Superior. Great cities stud its shore, but they are powerless to dim the transparency of its waters. Steamships cover the broad bosom of its lakes and estuaries; but they change not the beauty of the water no more than the fleets

of the world mark the waves of the oceans." (William Francis Butler. *The Great North Land*. 1872).

3.  Sir John Harington invented the principle of the water closet (as it is called in Europe) or flush toilet (in the American vernacular) in 1596. His invention did not come into extensive use until the early part of the 19th century.

4.  Water was once sold on the streets of Europe and North America, as it still is in the older sections of many Asian and African cities. A modern form of the water seller has been reappearing in most of the world's larger cities. Stocks of companies engaged in bottled-water distribution in urban centres are on the highly recommended lists of stockbrokers for active growth and development.

### 3 | Lakes are Made of Water

1.  The electromagnetic spectrum is commonly subdivided into wavelength or frequency bands in much the same way that radio frequencies are subdivided into commercial, police, and citizen bands.

2.  People who do not realize that water is more transparent to ultraviolet radiation at long wavelengths than it is to infrared (heat) radiation may develop intense sunburn on days when a light mist pervades the sky. Not feeling the direct heat of the sun's rays, they assume, erroneously, that the sun's radiation is evenly reduced over all wavelengths of the spectrum.

3.  For a description of this and other home experiments, see Vallentyne (1967).

4.  The two most common causes of this condition are (*i*) subsurface influx of salt-laden groundwater and (*ii*) extreme protection from the wind. Typically, both requirements have to occur simultaneously.

5.  *Seiche is* pronounced as in "saysh." The origin is probably from the Swiss-French word, *seiche,* meaning "sinking" (of water). This refers to the lowering of water level on the downwind end of a lake after a prolonged wind. The term is also said to have originated from the French word *sèche,* meaning dry.

### 4 | How Lakes Breathe

1.  Respiration in this book refers to the metabolic breakdown of organic substances in the cells of living organisms. Exchange of gases via the lungs is referred to as breathing.

2.  By a 1968 agreement between the governments of Canada and Ontario, 46 small headwater lakes and their land drainage basins were set aside for a period of 20 years for research on man-made eutrophication and related pollution problems in fresh water. Since that time, the initial agreement has been renegotiated on two occasions, each time expanding the list of management problems that can be the subject of experiments and increasing the number of lakes in the Experimental Lakes Area (ELA) to 58. The lakes are located in an isolated area in northwestern Ontario, just east of Lake of the Woods. This imaginative plan

was developed in 1965 by Dr. W.E. Johnson of the Fisheries Research Board of Canada (now a part of the Canadian Department of Fisheries and Oceans) and has been executed by staff from Canada's Freshwater Institute, in collaboration with other agencies and universities. For a general description of the ELA, see the *Journal of the Fisheries Research Board of Canada*, 1971, 28: 121 301 and 1973, 30: 1409 1552. Key studies of eutrophication in the ELA are described in Chapter 9.

3. Most of the decomposition of organic matter that takes place in lakes occurs in the epilimnion where life is abundant and active at higher temperatures than exist in the hypolimnion. The significant feature about the hypolimnion is the *very high ratio of decomposition to photosynthesis.*

4. The Latin name for iron is *ferrum*. Iron in ionic form can reversibly change from a ferric ($Fe^{3+}$) to a ferrous ($Fe^{2+}$) state depending on pH and concentration of dissolved oxygen. In the presence of high concentrations of dissolved oxygen, iron occurs as rust-coloured ferric oxides; in reduced (ferrous) form in sediments it often interacts with hydrogen sulfide to form a black precipitate of ferrous sulfide (FeS). It is often possible to guess the form in which iron occurs in sediments from colour alone.

## 5 | Phosphorus, the Morning Star

1. Antoine Lavoisier served the government of France on various commissions concerned with agriculture, saltpeter, and gunpowder; conditions in prisons and hospitals; education; and weights and measures. (The last-named led to the development of the metric system.) He was also an elected member of the Provincial Assembly at Orléans. Lavoisier's arrest during the French Revolution arose through previous financial association with the Ferme Générale, a tax collection agency. After 5 months of imprisonment, during which time no charges were laid, Lavoisier and 27 other members or former members of the Ferme Générale were—within the short span of 24 hours—charged with conspiracy, tried, found guilty by a Revolutionary Tribunal, and guillotined on May 8, 1794. An associate remarked the next day, "It required only a moment to sever that head, and perhaps a century will not suffice to produce another like it." Lavoisier's effects, confiscated by the French government at the time of his death, were returned 2 years later addressed to "The widow of the unjustly condemned Lavoisier."

2. Some organic compounds with phosphorus–carbon linkages are known. It has been suggested that phosphine ($PH_3$), a gas that spontaneously ignites in air, may be the cause of the flickering glow of the will-o'-the-wisp or *ignis fatuus* sometimes claimed to be seen on surfaces of marshes on pitch black nights. Although this is pure conjecture, it is not without the thread of possi-

bility. Some bacteria can transform phosphate into phosphine under laboratory conditions.

3.  Apatites occur as skeletal materials in vertebrates (as bone and in the enamel of teeth), in shells of marine phosphatic brachiopods, and in a few other groups.

## 6 | The Environmental Physician

1.  Lungs are organs of excretion, like the kidneys. An average adult exhales 0.25 litres (about 0.5 pints) of breath containing 5% carbon dioxide 20 times per minute. In 24 hours, this amounts to 0.67 kilogram (1.5 pounds) of carbon dioxide.

2.  The principal form of inorganic nitrogen in sewage is ammonia.

3.  Dr. W.T. Edmondson's work was supported by the U.S. National Science Foundation, a federal organization devoted to the support of science for cultural, rather than economic, purposes.

4.  In the 1970s, it seemed unlikely that household wastes from isolated homes and small communities would be treated for removal of phosphates because of poor control and high cost of treatment on a small scale. In North America, many summer cottages are equipped with all the luxuries of the city, including automatic dish washing and clothes washing machines. However, since that time, it has become common for dense cottage developments to have sewage systems with tertiary treatment, or tanks from which wastes are pumped and transported outside the lakes' basins. Another development becoming more widespread is the use of composting toilets, which release no wastes to water. Once composting is complete, only small amounts of solid and liquid effluents remain, which must then be disposed of carefully. In Europe, summer cottages are valued more for the contrast that they offer to city life. If so, nutrient pollution may be much less on a per capita basis.

## 7 | Detergents and Lakes

1.  Report to the International Joint Commission on the Pollution of Lake Erie, Lake Ontario, and the International Section of the St. Lawrence River. Vol. 1, Summary (1969); Vol. 2, Lake Erie (1970); Vol. 3, Lake Ontario and the International Section of the St. Lawrence River (1970). Ottawa and Washington.

2.  See U.S. House of Representatives (1970).

3.  Persistent foams from synthetic surfactants were acute problems in many areas of North America and other parts of the world during the late 1950s and early 1960s. The problem was solved in 1965 by the introduction of biodegradable surfactants.

4.  See Legge and Dingeldein (1970).

5.  See Kuentzel (1969); Lange (1967); and Kerr et al. (1970).

6. Nutrient deficiencies in nutrient-polluted lakes can be very misleading and should never be used as a basis for control measures. The important facts are what the limiting nutrient was *before* nutrient pollution, and the extent to which the total supply to water can be controlled by man.

7. See U.S. House of Representatives (1970).

8. All governments have wisely been extremely cautious in giving even the slightest suggestion of approval to any replacement for detergent phosphates. This is particularly true of the United Kingdom, many European countries, and a number of areas in the United States, where the recycling of water because of high population density can be appreciable in some areas; it is not uncommon to find 3–30% of ordinary tap water is derived from domestic and industrial wastes upstream. Under such conditions, toxicity and biodegradability become of paramount importance.

## 8 | The Year of NTA

1. Since STP breaks down (hydrolyzes) in the presence of water, its use is restricted to solid detergent products. NTA does not readily hydrolyze in water and can be used in liquid products.

2. The rationale for these first environmental studies on NTA was based on fish protection. Perhaps environmental benefit should be a basis for selecting components of detergent formulations.

3. STP was not put through all the tests that NTA had to pass. If the situation had been reversed, it is doubtful if STP would have survived the screening.

4. Canada had already enacted a regulation under the Canada Water Act, limiting the phosphate content of heavy-duty laundry detergents to a maximum of 20% as $P_2O_5$ (8.7% as P) after August 1, 1970. This was the first regulatory action taken in North America to limit the phosphate content of detergents.

5. Demotechnic refers to the combined effects of human population and technological production or consumption. The concept of a demotechnic explosion expresses more exactly than population explosion the revolution that has taken place in the past 200 years. See section on Terms and Definitions. Vallentyne (1972) coined the word "demophoric" and later (Vallentyne 1988) changed the word to "demotechnic."

6. This is referred to as the "cocktail" principle in Canadian environmental circles.

## 10 | Changes in the Eutrophication Problem Since the Mid-20th Century

1. The maps can be found on the Statistics Canada website: http://www.statcan.ca/english/freepub/16F0025XIB/m/manure.htm.

## 14 | The Eutrophication Problem as Part of a Greater Environmental Crisis

1.   See Jephson (1907).

2.   Jephson recorded there were some who objected: merchants and members of the municipal and county councils who did not want to pay higher taxes.

3.   Beard, D.P. 1996. Hard lessons from the US dam-building era. World Rivers Review 11 no. 4 (September) (http://internationalrivers.org/files/WRR.V11.N4.pdf).

# Glossary

**Algae:** Primitive photosynthetic plants that occur as microscopic forms suspended in water (phytoplankton), and as unicellular and filamentous forms attached to rocks and other substrates. About 15,000 species of freshwater algae are known.

**Alkalinity:** The buffering capacity of a lake, in general terms the lake's ability to neutralize strong inorganic acids such as sulphuric or hydrochloric acid.

**Allochthonous:** Having an origin outside a lake. For example, allochthonous organic carbon is usually that formed by photosynthesis in a lake's basin and washed or blown into the lake (compare with *Autochthonous*).

**Ammonia:** A nitrogen compound that occurs as a gas ($NH_3$) or, when dissolved in water, as an ion ($NH_4^+$).

**Anoxic:** Free of oxygen.

**Autochthonous:** Formed within a lake; for example, autochthonous carbon is that originating from photosynthesis of aquatic plants, including algae (compare with *Allochthonous*).

**Benthic shunt:** The process by which benthic molluscs, such as zebra mussels, filter algae from the water column, but release nutrients by excretion into the layer near a lake's bottom.

**Benthos:** The community of organisms living in contact with the bottom of a lake often subdivided into phytobenthos (plants) and zoobenthos (animals).

**Biomanipulation:** Manipulating the structure of a lake's food chain to achieve a management objective.

**Biosphere:** The region of the Earth inhabited by living organisms. The biosphere includes the lower part of the atmosphere, the entire hydrosphere (lakes, rivers, and oceans), and the lithosphere (soil and rock) to a depth of several kilometres.

**Bloom or water bloom:** A sudden increase in the abundance of planktonic algae, especially at or near the water surface. A condition when water looks green because of the abundance of planktonic algae.

**Blue-green algae:** The common name for members of the Cyanobacteria, a phylum of bacteria that obtain energy from photosynthesis. Some species contain pseudovacuoles that allow them to float near the surface. Others can fix nitrogen from the atmosphere, allowing them to thrive in situations that are common in many eutrophic lakes where dissolved phosphorus is abundant but dissolved ionic nitrogen is not.

**BOD:** Biological or biochemical oxygen demand; the amount of oxygen in milligrams per litre consumed in the biological decomposition of organic matter initially present in an enclosed sample of water. $BOD_5$ is the biochemical oxygen demand over a period of 5 days at $20°C$.

**Buffer:** A chemical that absorbs hydrogen ions and hydroxyl ions, thereby stabilizing solutions against dramatic changes in pH on addition of an acid or base.

**Buffering capacity:** The ability of a lake's water to neutralize strong inorganic acids.

**CAFO:** Confined animal feeding operation, or a large feedlot.

**Carbon:** A chemical element that occurs in inorganic (nonburnable) form or in the form of organic (burnable) compounds such as fats, oils, carbohydrates, and proteins.

**Catchment:** The terrestrial area that drains to a lake. In North America, the word watershed is usually used instead, but a watershed in the United Kingdom is the high point between two catchments.

**Celsius:** The temperature scale of the metric system in which freezing and boiling points of water at sea level are $0°C$ and $100°C$, respectively.

**Cladocera:** An order of crustaceans that is usually well represented in the zooplankton of lakes. Most species feed almost exclusively on algae.

**Cladoceran:** Belonging to the Cladocera.

**Copepoda:** An order of crustaceans that is usually well represented in the zooplankton of lakes. Several species may be present, and there are 11 juvenile stages before the animals become adults. Various species and larval stages are herbivores, omnivores, or carnivores. Carnivorous species feed on rotifers, smaller crustaceans, and protozoans.

**Cyanobacteria:** See *Blue-green algae*.

**Demophoric:** A term referring to human population and technological production considered jointly.

**DIC:** Dissolved inorganic carbon; the sum of carbon dioxide, bicarbonate ions, carbonate ions, and other forms of inorganic carbon dissolved in lakes.

**DOC:** Dissolved organic carbon; the sum of all forms of dissolved organic matter in lakes. Some is the result of excretion by organisms and decomposing aquatic plants, referred to as autochthonous DOC. Another source is the decomposition of terrestrial and wetland plants, which reaches lakes via streams, groundwater and carried by the wind, it is called allochthonous (see *Allochthonous*).

**DOM:** Dissolved organic matter; in most instances, it can be regarded as synonymous with DOC.

**Ecosystem:** Any combination of living and nonliving components that, with a supply of matter and energy, is self-sustaining over a defined period of time. An ecosystem can be an ocean, lake, small plot of land, the entire biosphere, or an aquarium, depending on the context of use.

**EDTA:** Ethylenediaminetetraacetic acid; a complexing agent similar to NTA.

**Efficiency:** The energy output of a process expressed as a percentage of the energy input.

**Enzymes:** Specific types of proteins formed in living cells that accelerate chemical reactions.

**Epilimnion:** (Plural, *epilimnia*; adjective, *epilimnetic*) the uniformly warm upper layer of a lake when it is thermally stratified in summer. The layer above the metalimnion.

**Eutrophic:** See *Lake classification*.

**Eutrophication:** The complex sequence of changes initiated by the enrichment of natural waters with plant nutrients. The first event in the sequence is an increased production and abundance of photosynthetic plants. This is followed by other changes that increase biological production at all levels of the food chain, including fish. Successional changes in species populations occur in the process. The original meaning of eutrophication was simply nutrient enrichment. In recent years, it has become more common to use the term in connection with the results rather than the cause (that is, an increase in trophic state caused by nutrient enrichment).

**Evapotranspiration:** The combined water loss from the Earth due to evaporation plus transpiration from plants.

**Flushing rate:** The number of times the entire water volume of a lake changes per year. For lakes with slow flushing, it can be a fraction, i.e. 0.01/year would be the rate for a lake with a 100 year flushing time.

**Flushing time:** The reciprocal of flushing rate, the number of years it would take to replace all of the water in a lake if it were emptied.

**Geochemistry:** The science dealing with chemical reactions that take place naturally on Earth.

**Halocline:** A density gradient caused by dissolved salts, rather than by temperature.

**Hardness:** The sum of calcium plus magnesium salts in water. An older term that is still widely used in water management.

**Hypereutrophic:** Extremely eutrophic (see *Lake classification*).

**Hypolimnion:** (Plural, *hypolimnia*; adjective, *hypolimnetic*) the uniformly cool and deep layer of a lake when it is thermally stratified in summer. The layer below the metalimnion.

**Ion:** An atom or molecule that, as a result of gaining or losing electrons, bears a positive or negative charge.

**Kilogram:** A unit of weight in the metric system, numerically equal to 2.2 pounds.

**Lake classification:** One of the more commonly used lake classification systems recognizes two general categories of lakes: dystrophic lakes with brown coloured water that is rich in humic materials derived from plants and oligotrophic—eutrophic lakes with "unstained" water.

*Oligotrophic lakes* are poorly supplied with plant nutrients and support little plant growth. As a result, biological productivity is generally low, the waters are clear, and the deepest layers are well supplied with oxygen through the year. Oligotrophic lakes tend to be deep, with average depths greater than 15 metres (49 feet) and maximum depths greater than 25 metres (80 feet).

*Mesotrophic lakes* are intermediate in characteristics between oligotrophic and eutrophic lakes. They are moderately well supplied with plant nutrients and support moderate plant growth.

*Eutrophic lakes* are richly supplied with plant nutrients and support heavy plant growths. As a result, biological productivity is generally high; the waters are turbid because of dense growths of phytoplankton, or contain an abundance of rooted aquatic plants; and the deepest waters exhibit reduced concentrations of dissolved oxygen during periods of restricted circulation. Eutrophic lakes tend to be shallow, with average depths less than 10 metres (33 feet) and maximum depths less than 15 metres (50 feet).

**Limnology:** The scientific study of inland waters.

**Litre:** A unit of volume in the metric system; approximately one Imperial quart.

**Meromixis:** A state where the deep waters of lakes are never completely mixed with surface waters because of higher salt concentrations near the bottom; these lakes are often in deep valleys that prevent the water from being mixed by the wind. The deep, unmixed waters of a meromictic lake are called the monimolimnion. The mixed surface waters are called the mixolimnion. The steep gradient between the two zones is known as the chemocline.

**Mesotrophic:** See *Lake classification.*

**Metalimnion:** (Plural, *metalimnia*; adjective, *metalimnetic*) the zone in which temperature decreases rapidly with depth in a lake when it is thermally stratified in summer. The metalimnion lies between the epilimnion and hypolimnion. The term is roughly equivalent to thermocline in ordinary usage.

**Microgram:** One thousandth of a milligram; one billionth of a kilogram.

**Milligram:** One thousandth of a gram; one millionth of a kilogram; 0.0000022 Imperial pound.

**Molar ratio:** The ratio of two chemicals, expressed in moles, where a mole is the atomic or molecular weight expressed in grams.

**Nitrate:** A negatively charged ion composed of one atom of nitrogen and three atoms of oxygen

**Nitrilotriacetic acid (NTA):** A complexing agent commonly sold in the form of its trisodium salt that is used to bind positively charged ions.

**Nitrogen (N):** A chemical element that occurs naturally in elemental form in air as nitrogen gas ($N_2$). Other inorganic forms of nitrogen in water are ammonium ($NH_4^+$), nitrite ($NO_2^-$), and nitrate ($NO_3^-$).

**Non-point sources:** Nutrients entering a waterbody from many points along its shore, for example, fertilizer seeping from fields.

**Oligotrophic:** See *Lake classification.*

**pH:** Negative logarithm of the concentration of hydrogen ions. A low pH means acidic conditions and a high concentration of hydrogen ions.

**Phosphate ($PO_4$):** A negatively charged ion composed of one atom of phosphorus and four atoms of oxygen.

**Phosphorus (P):** A chemical element. When used alone this term refers to the element in any chemical form.

**Photosynthesis:** The process by which green plants convert the sun's energy into carbohydrates.

**Phytoplankton:** Plant plankton (see *Plankton*).

**Piscivorous:** Fish-eating; examples are fish that eat other fish, such as northern pike and lake trout, or piscivorous birds, such as cormorants.

**Plankton:** Community of microorganisms, consisting of plants (*phytoplankton*) and animals (*zooplankton*), inhabiting open-water regions of lakes and rivers.

**Planktivorous:** Plankton-eating; there are zooplanktivorous species that eat zooplankton, such as minnows and larval fish, and phytoplanktivorous species like zooplankton, which feed on algae or other plants.

**Point sources:** Nutrients entering a waterbody at one point, such as an effluent pipe from a sewage treatment plant.

**Polymictic lake:** A lake that stratifies in calm weather, then mixes by wind several times a year.

**Primary production:** New organic material produced by photosynthesizing plants.

**Respiration:** The processes of enzymatic breakdown of organic substances in living cells that release energy for various biological activities.

**Rotifera:** Another common group in the zooplankton of lakes, belonging to the Phylum Rotatoria. Several species are usually present, and most feed on small algae and bacteria.

**Secchi depth:** Depth at which a Secchi disc (a small white disc, 20 centimetres or 8 inches in diameter) disappears from view when lowered into water. A measure of water transparency.

**Seiche:** (Pronounced *saysh*) oscillation of the surface of a lake (surface seiche) or interface between water layers of different density (internal seiche) with periodic times generally ranging from minutes to hours. Seiche movements are commonly initiated by prolonged unidirectional winds or gradients in barometric pressure.

**Supported lead:** The portion of the radioactive lead-210 in lake sediments that can be attributed to the radioactive decay of radium within the lake.

**Thermocline:** Literally, thermal or temperature gradient in a thermally stratified lake in summer; occupying the zone between the epilimnion and hypolimnion and more or less equivalent to the term metalimnion.

**Trophic state:** Characterization of a body of water in terms of position in a scale ranging from oligotrophy to eutrophy.

**Trophogenic zone:** Nourishment-producing; the upper well-illuminated region of a waterbody in which photosynthesis predominates. The nutrients produced by the photosynthetic organisms then pass up the food chain.

**Unsupported lead:** The portion of the radioactive lead-210 in lake sediments that results from atmospheric fallout.

**Varves:** Distinct, annual laminations in lake sediments, generally visible as paired light and dark bands.

**Zooplankton:** (Pronounced as in *zoology*) Animal plankton (see *Plankton*).

# References

Anneville, O., and Pelletier, J.P. 2000. Recovery of Lake Geneva from eutrophication: quantitative response of the phytoplankton. Arch. Hydrobiol. 148: 607–624.

Beard, D.P. 1996. Hard lessons from the US dam-building era. World Rivers Review 11 no. 4-September. Available from http://internationalrivers.org/files/WRR. V11.N4.pdf.

Birge, E.A. 1910. On the evidence for temperature seiches. Trans. Wis. Acad. Sci. Arts Lett.16: 1005–1016.

Birge, E.A. 1916. The work of the wind in warming a lake. Trans. Wis. Acad. Sci. Arts Lett. 18: 341–391.

Boesch, D., Hecky, R.E., O'Melia, C., Schindler, D.W., and Seitzinger, S. 2006. Eutrophication of Swedish seas. Naturvardverket, Stockholm. 72 pp.

Bonnifield, P. 1979. The dust bowl: men, dirt, and depression. University of New Mexico Press, Albuquerque. 232 pp.

Bossard, P., and Bürgi, H. 2005. Eutrophication and re-oligotrophication of Lake Lucerne. Available from http://homepages.eawag.ch/~bossard/projektb.html.

Bowman, M.F., Chambers, P.A., and Schindler, D.W. 2007. Constraints on benthic algal response to nutrient addition in oligotrophic mountain rivers. River Res. Appl. 23: 858–876.

Brattberg, G. 1986. Decreased phosphorus loading changes phytoplankton composition and biomass in the Stockholm archipelago. Vatten, 42: 141–152.

Broecker, W.S. 1974. Chemical oceanography. Harcourt Brace Jovanovich, Inc., New York.

Brown, L.R. 2006. Plan B 2.0: rescuing a planet under stress and a civilization in trouble. W.W. Norton & Co., New York. 352 pp.

Bührer, H., and Ambühl, H. 2001. Lake Lucerne, Switzerland, a long term study of 1961–1992. Aquat. Sci. 63: 432–456.

Carpenter, S.R. 2003. Regime shifts in lake ecosystems: pattern and variation. Excellence in ecology series vol. 15. International Ecology Institute, Oldendorf/Luhe, Germany. 199 pp.

Carpenter, S.R. 2005. Eutrophication of aquatic ecosystems: biostability and soil phosphorus. Proc. Natl. Acad. Sci. USA.102: 10002–10005.

Carpenter, S.R., Kitchell, J.F., and Hodgson, J.R. 1985. Cascading trophic interactions and lake productivity. BioScience, 35: 634–639.

Carson, R. 1962. Silent spring. Fawcett Crest, New York, 304 pp.

Chambers, P.A., Guy, M., Roberts E.S., Charlton, M.N., Kent, R., Gagnon, C., et al. 2001. Nutrients and their impact on the Canadian environment. Agriculture and Agri-Food Canada, Environment Canada, Fisheries and Oceans Canada, Health Canada and Natural Resources Canada, Ottawa, ON.

Chamut, P.S., Pond, S.G., and Taylor, V.R. 1972. The "red herring" pollution crisis in Placentia Bay, Newfoundland. A general description and chronology. Fisheries Research Board of Canada, Ottawa, ON. Circ. No. 2. pp. 29–43.

Chan, F., Marino, R.L., Howarth, R.W., and Pace, M.L. 2006. Ecological constraints on planktonic nitrogen fixation in saline estuaries. II. Grazing controls on cyanobacterial population dynamics. Mar. Ecol. Prog. Ser. 309: 41–53.

Chapra, S.C. 1977. Total phosphorus model for the Great Lakes. J. Environ. Eng. Div. 103: 153.

Charleton, M.N., and Milne, J.E. 2004. Review of thirty years of change in Lake Erie water quality. National Water Research Institute, Burlington, ON. NWRI Contrib. No. 04–167.

Cole, J.J., Caraco, N.F., Kling, G.W., and Kratz, T.K. 1994. Carbon dioxide super-saturation in the surface waters of lakes. Science (Washington, DC), 265: 1568–1570.

Commoner, B. 1971. The closing circle. Knopf, New York. 326 pp.

Cooke, D.G., Welch, E.B., Peterson, S., and Nichols, S.A. 2005. Restoration and management of lakes and reservoirs. 3rd ed. CRC Press, Boca Raton, FL. 591 pp.

Daily, G.C., and Ellison, K. 2002. The new economy of nature: the quest to make conservation profitable. Island Press, Washington, DC. 260 pp.

Dillon, P.J., and Rigler, F.H. 1974. A test of a simple nutrient budget model predicting the phosphorus concentration in lake water. J. Fish. Res. Board Can. 31: 1771–1778.

Dise, N., and Wright, R.F. 1995. Nitrogen leaching from European forests in relation to nitrogen deposition. For. Ecol. Manage. 71: 153–162.

Douglas, M.S.V., Smol, J.P., Savelle, J.M., and Blais, J.M. 2004. Prehistoric Inuit whalers affected Arctic freshwater ecosystems. Proc. Natl. Acad. Sci. USA. 101: 1613–1617.

Eckhéll, J., Jonsson, P., Meili, M., and Carman, R. 2000. Storm influence on the accumulation and lamination of sediments in deep areas of the northwestern Baltic proper. Ambio, 29: 238–245.

Edmondson, W.T. 1991. The uses of ecology: Lake Washington and beyond. University of Washington Press, Seattle. 329 pp.

Elser, J.J., Sterner, R.W., Galford, A.E., Chrzanowski, T.H., Findlay, D.L., Mills, K.H., et al. 2000. Pelagic C:N:P stoichiometry in a eutrophied lake: responses to a whole-lake food-web manipulation. Ecosystems, 3: 293–307.

Elton, C.S. 1958. The ecology of invasions by animals and plants. Methuen & Co. Ltd., London. 181 pp.

Emerson, S. 1975a. Chemically enhanced $CO_2$ gas exchange in a eutrophic lake: a general model. Limnol. Oceanogr. 20: 743–753.

Emerson, S. 1975b. Gas exchange rates in small Canadian Shield lakes. Limnol. Oceanogr. 20: 754–761.

Fixen, P.E., and West, F.B. 2002. Nitrogen fertilizers: meeting contemporary challenges. Ambio, 31: 169–176.

Flett, R.J., Schindler, D.W., Hamilton, R.D., and Campbell, N.E.R. 1980. Nitrogen fixation in Canadian Precambrian Shield lakes. Can. J. Fish. Aquat. Sci. 37: 494–505.

Fonselius, S.H., and Valderrama, J. 2003. One hundred years of hydrographic measurements in the Baltic Sea. J. Sea Res. 49: 229–241.

Forbes, S.T. 1887. The lake as a microcosm. Bull. Peoria Ill. Sci. Assoc. 1887: 77–87.

Granat, L. 2001. Deposition of nitrate and ammonium from the atmosphere to the Baltic Sea. In Wulff, F., Rahm, L., and Larsson, P. (eds). A systems analysis of the Baltic Sea. Springer-Verlag, Berlin. pp. 133–148.

Hecky, R.E. 2004. The nearshore phosphorus shunt: a consequence of ecosystem engineering by dreissenids in the Laurentian Great Lakes. Can. J. Fish. Aquat. Sci. 61: 1285–1293.

Hecky, R.E., and Hesslein, R.H. 1995. Contributions of benthic algae to lake food webs as revealed by stable isotope analysis. J. N. Am. Benthol. Soc. 14: 631–653.

Hodder, V.M., Parsons, L.S., and Pippy, J.H.C. 1972. The occurrence and distribution of "red" herring in Placentia Bay, February–April 1969. Fisheries Research Board of Canada, Ottawa, ON. Circ. No. 2. pp. 45–52.

Howarth, R.W. 1998. An assessment of human influences on fluxes of nitrogen from the terrestrial landscape to the estuaries and continental shelves of the North Atlantic Ocean. Nutrient Cycling Agroecosyst. 52: 213–223.

Howarth, R.W., Chan, F., and Marion, R. 1999. Do top-down and bottom-up controls interact to exclude nitrogen-fixing cyanobacteria from the plankton of estuaries? An exploration with a simulation model. Biogeochemistry, 46: 203-231.

Hrudey, S.E., and Hrudey, E.J. 2004. Safe drinking water: lessons from recent outbreaks in affluent nations. IWA Publishing, Cornwall, UK. 486 pp.

Hutchinson, G.E. 1957. A treatise on limnology. Vol. 1. John Wiley & Sons, New York. 1015 pp.

Hutchinson, G.E., 1969. Eutrophication, past and present. In Rohlich, C.A. (ed). Eutrophication, causes, consequences, correctives. National Academy of Sciences, Washington, DC. pp. 17-26.

Hutchinson, G.E. 1973. Eutrophication. Am. Sci. 61: 269-279.

Hutchinson, G.E., Bonatti, E, Cowgill, U.M., Goulden, C.E., Leventhal, E.A., Mallett, M.E., et al. 1970. Ianula: an account of the history and development of the Lago di Monterosi, Latium, Italy. Trans. Amer. Phil. Soc. 60: 1-178.

Jangaard, P.M. 1970. The role played by the Fisheries Research Board of Canada in the "red" herring phosphorus pollution crisis in Placentia N. Bay, Newfoundland. Fisheries Research Board of Canada, Office of the Atlantic Regional Director, Halifax, NS. Circ. No. 1. pp 7-26.

Jephson, H. 1907. The sanitary evolution of London. T. Fisher Unwin, London. 440 pp.

Jeppesen, E., Søndegarard, M., Jensen, J.P., Havens, K.E., Anneville, O., Carvalho,, L., et al. 2005. Lake responses to reduced nutrient loading—an analysis of contemporary long-term data from 35 case studies. Freshwater Biol. 50: 1747-1771.

Johnson, W.E., and Vallentyne, J.R. 1971. Rationale, background, and development of experimental lake studies in northwestern Ontario. J. Fish. Res. Board. Can. 28: 123-128.

Jonsson, P., Persson, J., and Holmberg, P. (eds). 2003. Skärgårdens bottnar : en sammanställning av sedimentundersökningar gjorda 1992-1999 i skärgårdsområden längs svenska ostkusten. Naturvårdsverket, Stockholm, Sweden. Rapp. No. 5212. 112 pp. [In Swedish.]

Kelly, C.A., Fee, E., Ramlal, P.S., Rudd, J.W.M., Hesslein, R.H., Anema, C., et al. 2001. Natural variability of carbon dixoide and net epilimnetic production in the surface waters of boreal lakes of different sizes. Limnol. Oceanogr. 46: 1054-1064.

Kerr, P.C., Paris, D.R., and Brockway, D.L. 1970. The interrelationship of carbon and phosphorus in regulating heterotrophic and autoitrophic populations in aquatic ecosystems. U.S. Department of the Interior, Federal Water Pollution Control Administration, Washington, DC. Water Pollut. Control Res. Ser. No. 16050 FGS 07/70.

Kidd, K.A., Blanchfield, P.J., Mills, K.H., Palace, V.P., Evans, R.E., Lozorchak, J.M., et al. 2007. Collapse of a fish population after exposure to a synthetic estrogen. Proc. Natl. Acad. Sci. USA 104: 8897-8901.

Kling, H.J., Findlay, D.L., and Komarek, J. 1994. *Aphanizomenon schindleri*: a new nostocacean cyanoprokaryote from the Experimental Lakes Area, northwestern Ontario. Can. J. Fish. Aquat. Sci. 51: 2267-2273.

Kuentzel, L.E. 1969. Bacteria, carbon dioxide and algal bloom. J. Water Pollut. Control Fed. 41: 1737-1747.

Laird, K.R., Cumming, B.F., Wunsam, S.R.J., Oglesby, R.J., Fritz, S.C., and Leavitt, P.R. 2003. Lake sediments record large-scale shifts in moisture regimes across the northern prairies of North America during the past two millennia. Proc. Natl. Acad. Sci. USA. 100: 2483-2488.

Lake Winnipeg Stewardship Board. 2006. Reducing nutrient loading to Lake Winnipeg and its watershed. Our collective responsibility and commitment to action. Report to the Manitoba Minister of Water Stewardship, Winnipeg. December 2006. Available from www.gov.mb.ca/waterstewardship/water_quality/lake_winnipeg/lwsb2007-12-final-rpt.pdf.

Lange, W. 1967. Effect of carbohydrates on the symbiotic growth of planktonic blue-green algae with bacteria. Nature (London), 215: 1277-1278.

Lange, W. 1970. Cyanophyta-bacteria systems: effects of added carbon compounds or phosphate on algal growth at low nutrient concentrations. J. Phycol. 6: 230-234.

Lännergren, C. 1994. Stockholms Skärgård. *In* Swedish Report on the Environmental State of the Baltic Proper. pp. 1-3. [In Swedish with English summary.]

Lännergren, C., and Eriksson, B. 2005. Undersökningar I Stocksholms Skärgård 2004. Stockholm Vatten MV-05110.

Larsson, U., Elmgren, R., and Wulff, F. 1985. Eutrophication and the Baltic Sea: causes and consequences. Ambio, 14: 9-14.

Legge, R.F., and Dingeldein, D. 1970. We hung the phosphates without a fair trial. Can. Res. Dev. 3: 19-42.

Likens, G.E. (ed). 1972. Nutrients and eutrophication: the limiting-nutrient controversy. Limnol. Oceanogr. Spec. Symp. Vol. 1. 328 pp.

Marino, R., Chan, F., Howarth, R.W., Pace, M.L., and Likens, G.E. 2006. Ecological constraints on planktonic nitrogen fixation in saline estuaries. I. Nutrient and trophic controls. Mar. Ecol. Prog. Ser. 309: 25-39.

Mortimer, C.H. 1953. The resonant response of stratified lakes to wind. Schweiz. Seits. Hydrol. 15: 94-151.

National Academy of Sciences. 1969. Eutrophication: Causes, Consequences, Correctives. International Symposium on Eutrophication, 1967, Madison, WI. U.S. National Academy of Sciences, Washington, DC.

Neufeld, S.D. 2005. Effects of catchment land use on nutrient export, stream water chemistry, and macroinvertebrate assemblages in boreal Alberta. M.Sc. thesis, University of Alberta, Edmonton.

Pace, M.L., Cole, J.C., Carpenter, S.R., Kitchell, J.F., Hodgson, J.R., Van de Bogert, M.C., et al. 2004. Whole-lake carbon-13 additions reveal terrestrial support of aquatic food webs. Nature (London), 427: 240–243.

Patalas, K. 1984. Mid-summer mixing depths of lakes of different latitudes. Verh. Int. Verein. Theor. Angew. Limnol. 22: 97–102.

Post, J.R., Sullivan, M., Cox, S., Lester, P., Walters, C.J., Parkinson, E.A., et al. 2002. Canada's recreational fisheries: the invisible collapse? Fisheries, 27: 6–17.

Sauchyn, D.J., and Skinner, W.R. 2001. A proxy record of drought severity for the southwestern Canadian plains. Can. Water Resour. J. 26: 253–272.

Sauchyn, D.J., Barrow, E., Hopkinson, R.F., and Leavitt, P. 2002. Aridity on the Canadian Plains. Géogr. Phys. Quat. 56: 247–259.

Sauchyn, D.J., Beriault, A., and Stroich, J. 2003. A paleoclimatic context for the drought of 1999–2001 in the northern Great Plains. Geogr. J. 169: 158–167.

Sauchyn, D., Pietroniro, A., and Demuth, M. 2006. Upland watershed management and global change—Canada's Rocky Mountains and western plains. Fifth Biennial Rosenberg Forum on Water Policy, September 6–11, 2006, Banff, AB.

Scheffer, M., Carpenter, S., Foley, J.A., Folke, C., and Walker, B. 2001. Catastrophic shifts in ecosystems. Nature (London), 413: 591–596.

Scheffer, M., Hosper, S.H., Meijer, M.-L., Moss, B., and Jeppesen, E. 1993. Alternative equilibrium in shallow lakes. Trends Ecol. Evol. 8: 275–279.

Schelske, C.L. 1999. Diatoms as mediators of biogeochemical silica depletion in the Laurentian Great Lakes. In Stoermer, E.F., and Smol, J.P (eds.). The diatoms: applications for the environmental and earth sciences. Cambridge University Press, Cambridge, UK. pp. 73–84.

Schelske, C.L., and Stoermer, E.F. 1971. Eutrophication silica depletion and predicted changes in algal quality in Lake Michigan. Science (Washington, DC), 173: 423–424.

Schindler, D.E., Geib, S.I., and Williams, M.R. 2000. Patterns of fish growth along a residential development gradient in north temperate lakes. Ecosystems, 3: 229–237.

Schindler, D.W. 1971a. Light, temperature and oxygen regimes of selected lakes in the Experimental Lakes Area (ELA), northwestern Ontario. J. Fish. Res. Board Can. 28: 157–170.

Schindler, D.W. 1971b. Carbon, nitrogen, phosphorus and the eutrophication of freshwater lakes. J. Phycol. 7: 321–329.

Schindler, D.W. 1974. Eutrophication and recovery in experimental lakes: implications for lake management. Science (Washington, DC) 184: 897–899.

Schindler, D.W. 1975. Whole-lake eutrophication experiments with phosphorus, nitrogen and carbon. Verh. Int. Verein. Theor. Angew. Limnol. 19: 3221–3231.

Schindler, D.W. 1977. Evolution of phosphorus limitation in lakes: natural mechanisms compensate for deficiencies of nitrogen and carbon in eutrophied lakes.
Science (Washington, DC), 195: 260-262.

Schindler, D.W. 1998. Replication versus realism: the need for ecosystem-scale
experiments. Ecosystems, 1: 323-334.

Schindler, D.W. 2001. The cumulative effects of climate warming and other human
stresses on Canadian freshwaters in the new millennium. Can. J. Fish. Aquat.
Sci. 58: 18-29.

Schindler, D.W., and Comita, G.W. 1972. The dependence of primary production
upon physical and chemical factors in a small senescing lake, including the
effects of complete winter oxygen depletion. Arch. Hydrobiol. 69: 413-451.

Schindler, D.W., and Fee, E.J. 1973. Diurnal variation of dissolved inorganic carbon
and its use in estimating primary production and $CO_2$ invasion in Lake 227. J.
Fish. Res. Board Can. 30: 1501-1510.

Schindler, D.W., Armstrong, F.A.J., Holmgren, S.K., and G.J. Brunskill. 1971.
Eutrophication of Lake 227, Experimental Lakes Area (ELA), northwestern
Ontario, by addition of phosphate and nitrate. J. Fish. Res. Board Can. 28:
1763-1782.

Schindler, D.W., Brunskill, G.J., Emerson, S., Broecker, W.S., and Peng, T.-H. 1972.
Atmospheric carbon dioxide: its role in maintaining phytoplankton standing
crops. Science (Washington, DC), 177: 1192-1194.

Schindler, D.W., Kling, H., Schmidt, R.V., Prokopowich, J., Frost, V.E., Reid R.A.
and M. Capel. 1973. Eutrophication of Lake 227 by addition of phosphate and
nitrate: the second, third and fourth years of enrichment 1970, 1971 and 1972. J.
Fish. Res. Board Can. 30: 1415-1440.

Schindler, D.W., Lean, D., and Fee, E. 1975. Nutrient cycling in freshwater ecosys-
tems. In Productivity of World Ecosystems. Proceedings of a Symposium,
August 31-September 1, 1972, Seattle, WA. National Academy of Sciences,
Washington, DC. pp. 96-105.

Schindler, D.W., Fee, E.J., and Ruszczynski, T. 1978. Phosphorus input and its consequences for phytoplankton standing crop and production in the Experimental
Lakes Area and in similar lakes. J. Fish. Res. Board Can. 35: 190-196.

Schindler, D.W., Hecky, R.E., and Mills, K.H. 1993. Two decades of whole lake
eutrophication and acidification experiments. In Rasmussen, L., Brydges, T.,
and Mathy, P. (eds). Experimental manipulations of biota and biogeochemical
cycling in ecosystems. Commission of the European Communities, Brussels.
pp. 294-304.

Schindler, D.W., Curtis, P.J., Parker, B., and Stainton, M.P. 1996a. Consequences of
climate warming and lake acidification for UV-B penetration in North American
boreal lakes. Nature (London), 379: 705-708.

Schindler, D.W., Bayley, S.E., Parker, B.R., Beaty, K.G., Cruikshank, D.R., Fee, E.J., et al. 1996b. The effects of climatic warming on the properties of boreal lakes and streams at the Experimental Lakes Area, northwestern Ontario. Limnol. Oceanogr. 41: 1004–1017.

Schindler, D.W., Anderson, A.-M., Brzustowski, J., Donahue, W.F., Goss, G., Nelson, J., et al. 2004. Lake Wabamun: a review of scientific studies and environmental impacts. Prepared for the Minister of Alberta Environment, Edmonton. Publ. No. T769.

Schindler, D.W., Dillon, P.J., and Schreier, H. 2006. A review of anthropogenic sources of nitrogen in Canada and their effects on Canadian aquatic ecosystems. Biogeochemistry, 79: 25–44.

Schindler, D.W., Wolfe, AP, Vinebrooke, R., Crowe, A., Blasi, J.M., Miskimmin, B., and Freed, R. 2008a. The cultural eutrophication of Lac la Biche, Alberta, Canada: a paleoecological study. Can. J. Fish. Aquat. Sci. 65(10). In press.

Schindler, D.W., Hecky, R.E., Findlay, D.L., Stainton, M.P., Parker, B.R., Paterson, M.J., Beaty, K.G., Lyng, M., and Kasian, S.E.M. 2008b. Eutrophication of lakes cannot be controlled by reducing nitrogen input: results of a 37-year whole-ecosystem experiment. Proc. Natl. Acad. Sci. USA, 105: 1254–11258.

Skjelvåle, B.L., Stoddard, J.L., Jeffries, D.S., Tørseth, K., Høgåsen, T., Bowman, J., et al. 2005. Regional scale evidence for improvements in surface water chemistry 1990–2001. Environ. Pollut. 137: 165–176.

Smol, J.P. 2008. Pollution of lakes and rivers: a paleoenvironmental perspective. 2nd ed. Blackwell Publishing, Oxford, UK.

Søndergaard, E., Jeppesen, J., Jensen, P., and Lauridsen, T. 2000. Lake restoration in Denmark. Lakes Reservoir Res. Manage. 5: 151–159.

Søndergaard, M., Jensen, J.P., Jeppesen, E., and Møller, P.H. 2002. Seasonal dynamics in the concentrations and retention of phosphorus in shallow Danish lakes after reduced loading. Aquat. Ecosyst. Health Manage. 5: 19–20.

Sullivan, M. 2003. Active management of walleye fisheries in Alberta: dilemmas of managing recovering fisheries. N. Am. J. Fish. Manage. 23: 1343–1358.

Tarranson, L., and Schaug, J. 2000. Transboundary acidification and eutrophication in Europe. EMEP summary report 2000. Norwegian Meteorological Institute, Oslo, Norway. Publ. No. EMEP/MSC-W 1/2000.

Union of Concerned Scientists. 2007. Smoke, mirrors & hot air: How ExxonMobil uses big tobacco's tactics to manufacture uncertainty on climate science. Available from www.ucsusa.org/assets/documents/global_warming/exxon_report.pdf.

U.S. House of Representatives. 1970. Phosphates in detergents and the eutrophication of America's waters. In Hearings before a Subcommittee of the Committee on Government Operations. House of Representatives, Ninety-first

Congress, first session, Dec. 15 and 16, 1969. U.S. Government Printing Office, Washington, DC.

Vallentyne, J.R. 1967. A simplified model of a lake for instructional use. J. Fish. Res. Board Can. 24: 2473–2479.

Vallentyne, J.R. 1972. Demophora. Environment, 14: 47–48.

Vallentyne, J.R. 1974. The algal bowl: lakes and man. Department of the Environment, Fisheries and Marine Service, Ottawa, ON. Misc. Spec. Publ. No. 22.

Vallentyne, J.R. 1988. First direction, then velocity. Ambio, 17: 409.

Vallentyne, J.R. 2006. Tragedy in mouse utopia: the sorcerer lurks within. An ecological commentary on human utopia. Trafford Publishing, Victoria, BC. 198 pp.

Vollenweider, R.A. 1968. Scientific fundamentals of the eutrophication of lakes and flowing waters, with particular reference to nitrogen and phosphorus as factors in eutrophication. Organisation for Economic Co-operation and Development, Paris. OECD Tech. Rep. No. DAS/CS/68.27.

Vollenweider, R.A. 1976. Advances in defining critical loading levels for phosphorus in lake eutrophication. Mem. Ist. Ital. Idrobiol. 33: 53–83.

Vollenweider, R.A., Marchetti, R., and Viviani, R. 1992. Marine coastal eutrophication. Elsevier Science Publishers B.V., Amsterdam.

Wackernagel, M. and Rees, W.E. 1996. Our ecological footprint: reducing human impact on the earth. New Society Publishers, Gabriola Island, BC. 160 pp.

Weber, C.A. 1907. Aufbau und Vegetation der Moore Norddeutschlands. Beibl. Bot. Jahrb. 90: 19–34.

Wilander, A., and Persson, G. 2001. Recovery from eutrophication: experiences of reduced phosphorus input to the four largest lakes of Sweden. Ambio, 30: 475–485.

Willen, E. 2001 Phytoplankton and water quality characterization: experiences from the Swedish large lakes Mälaren, Hjälmaren, Vättern and Vänern. Ambio, 30: 529–537.

Wright, R.F., Alewell, C., Cullen, J.M., Evans, C,D., Marchetto, A., Moldan, F., et al. 2001. Trends in nitrogen deposision and leaching in acid-sensitive streams of Europe. Hydrol. Earth Syst. Sci. 5: 299–310.

Wulff, F., Rahm, L., and Larsson, P. 2001. Introduction. *In* Wulff, F.V., Rahm, L.A., and Larsson, P. (eds). A systems analysis of the Baltic Sea. Ecol. Stud. 148. pp. 1–17.

# Index

builder (detergent), 126–28, 133, 134, 139, 144

Bull Run, 210

Bureau of Reclamation (U.S.), 274, 275, 277

*Bythothrephes*, 76

caddisflies, 52

cadmium, 146, 197

caffeine, 198, 229

CAFO, *see* confined animal feeding operation

calcareous, 8, 22, 93

calcium, 24, 69, 71, 72, **74**, 93, 109, 111, 112, 126, 127, 138

calcium carbonate, 20, 72–75, 112, 133

California, 59

*Campylobacter jejuni*, 193, 206

Canada Centre for Inland Waters, 145, 163

Canada Water Act, 291n.8.4

canals, 107, 108, 260, 269

cancer, 125, 145, 149, 153

canthaxanthin, 224, **229**

carbohydrates, 67, 92

carbon, 13, 38, 67–70, 73, **74**, 78, 90, 110, 123, **124**, 125, 157–60, 161, **162**, 165–67, **168**, 169, 180, 215, 216, 223, 244, 289n.5.2

carbon dioxide, 3, 13, 38, 64, 66–68, 70, 72–75, 81–85, 92, 99, 124, 159, 160–63, 165, 166, 169, 170, 260, 290n.6.1

carbonate, 69–73, **74**, 93, 132, 141

carcinogen, 40, 144, 145, 150

carnivore, 77, 80

carp (*Cyprinus carpio*), 25, 26, 114, 199

cars, *see* automobiles

catch per unit effort, 174

catchment, x, xi, 3, 9, 10, 12, 22, 24–27, 32, 33, 35, 41, 70, 78, 80, 82, 156, 158, 170, 180, 187, 190, 193–95, 200, **207**, 209, 211, 212, 214, 216, **219**, 224, 231, 235, 239, **243**, 263, 268, 272

cation, 71, 156, 180

Cascade Range, 210

Catastrophe Bay, **211**

Catskill Mountains, 209

cattle, 3, 15, 65, 80, 192, 193, 195, **208**, 209, 264

CCIW, *see* Canada Centre for Inland Waters

cesium, 46, 215

cesspool, 270

*Chaoborus*, 76, 217

Char Lake, **163**

Chesapeake Bay, 106, 246

chinampas, 107

Chironomidae, 217

*Chironomus plumosus*, 5, 6, **7**

chlorine, 20, 30, 40, 102, 153, 209

chlorophyll, **125**, 253, 254

chlorophyll a, 125, 157, 171, **172**, 224, **229**, 237, 238, 240, **254**

cholera, 12, 37, 102, 108, 111, 269

chromium, 197

chronic feeding study, 144

chronic toxicity, 144

Chrysophycea, 1, 217, 224, **227**

Churchill River, 218

*Chydorus*, **230**

cisco (*Coregonus artedi*), 3, 4, 27, 29, 52, 220

cistern, 271, 279

citric acid, 141

Cladocera, 26, 76, **79**, 80, 174, 217, 223, **230**

*Cladophora*, 20, 32

Clean Water Act, x, xi, 176, 177, 265, 273

Clear Lake, 142

climate warming, ix, x, 10, 68, 103,
   161, 165, 179, 182–85, 187–89, **207**,
   246, 267
coal, 28, 64, 67, 216
coarse woody debris, 204, **205**
coliform bacteria, 11, 103, 176, 223
Committee on Government Operations
   (U.S.), 121
composting, 114, 195, 198, 229, 263,
   278, 280, 290n.6.4
confined animal feeding operation, 110,
   189–94, **208**, 243, 263
contaminants, 180, 187, 192, 206, 209,
   273, 274, 277
Copepoda, **79**, 80, 173, 200, 217
copper, 24, 25, 104, 105, 145, 197
cormorant (*Phalacrocorax auritus*),
   202, 231
cosmetics, 113, 139, 153
cottage development, 180, 181, 185,
   205–07, 220, 221, 228, 237,
   262–64, 290n.6.4
Council on Environmental Quality
   (U.S.), 147
cows, *see* cattle
*Craspedacusta*, 76
Crater Lake, 2, 59, 81
cropland, 9, 42, 85, 108, 112, 114, 184,
   189, 190, **196**, 228, 237, 243, 275
crustaceans, 9, 53, 76, 79, 80, 84, 200
cryptophytes, 1
*Cryptosporidium*, 11, 40, 102, 206, 209
*Cryptosporidium parvum*, 193
cultural eutrophication, 2, 5, 9, 12, 37,
   48, 85, 88, 99, 100, 101, 109–11, 120,
   123, 128, 177, 233
Cuyahoga River, 31
cyanide, 97
Cyanobacteria, 2, 7, **11**, 21, 29, **31**, 34,
   67, 94, 167–69, 174–76, 189, 210,

224, 228, **230**, 245–49, 251, 253,
   256
*Cyclotella*, 224, 228
*Cyclotella bodanica*, **228**

2,4-D, *see* 2,4-dichlorophenoxyacetic
   acid
dam, 24, 35, 44, 204, 237, 274–77,
   292n.14.3
*Daphnia*, **79**, 80, 173–75, 201, **207**,
   217, **230**
*Daphnia galeata mendotae*, 27, 173, 174
*Daphnia pulicaria*, 27, 174
DDT, *see* dichloro-diphenyl-
   trichloroethane
dead zone, 4, 241, 247
decomposition, 2, 16, 30, 39, 75–78,
   80–82, 93, 114, 142, 145, 150, 183,
   192, 245, 249, 256, 289n.4.3
Delta Marsh, **211**
demophoric growth, 291n.8.5
demotechnic growth, 151, 152, 261,
   262, 291n.8.5
denitrification, 112, 196, 234, 246
Denmark, 235, 239, **247**
deoxyribonucleic acid, 90, 91
Department of the Environment
   (Canada), 148, 149, 200, 212
Department of the Interior (Canada),
   220
destratification, 105, 234
detergent, 10, 14, 31, 95, 99, 100, 118,
   119–35, 139–41, 143, 144, 146–49,
   151, 153, 157, 165, 180, **181**, 250,
   267, 291n.8.1, 291n.8.2, 291n.8.4
detritus, 78
Detroit River, 287n.2.1
Devils Lake, 35
diarrhea, 206, 269
diatoms, 7, 14, 20, 21, 29, 188, 189,
   216, 217, 223, 224, **227**, **228**, 230

diatoxanthin, 224, **229**

DIC, *see* dissolved inorganic carbon

dichloro-diphenyl-trichloroethane, 131

2,4-dichlorophenoxyacetic acid, 105

dimictic lakes, 240

dinoflagellates, 1

dirty thirties, xi, 188

disease, 11, 12, 16, 37, 38, 44, 100, 102, 107, 111, 206, 209, 211, 269–72

dissolved inorganic carbon, **83**, 157–61, 165

dissolved organic carbon, 55–57, 70

dissolved organic matter, 20, 70, 78

DNA, *see* deoxyribonucleic acid

DOC, *see* dissolved organic carbon

DOM, *see* dissolved organic matter

drought, x, 35, 182, 184, 187, 188, 193

Dust Bowl, x, xi, 10

East China Sea, 242

Echimamish River, 33

echinone, 224, **229**

ecological footprint, 261

ecology, 77, 261

ecosystem, x, xi, 27, 42, 52, 65, 68, 77, 78, 80, 81, 85, 101, 102, 105, 108, 110, 139, 150, 166, 196, 197, 200, 212, 236, 241, 244, 249, 263, 266, 274, 276, 280

EDTA, *see* ethylenediaminetetraacetic acid

eelgrass, 247

effluent, 4, 10, 14, 15, 24, 25, 35, 41, 97, 100, 104, 106, 107, 109, 113, 114, 116, 117, 120, 131, 167, 180, 187, 197, 198, 223, 251, 256, 290n.6.4

ELA, *see* Experimental Lakes Area

Elk Island, **211**

Elwha River, 277

England, 82

enrichment, 9, 10, 19, 21, 73, 106, 131, 165, 177, 256

Environment Canada, 264

Environmental Protection Agency (U.S.), xi, 147, 176, 177, 209, 264, 266

EPA, *see* Environmental Protection Agency

epilimnion, 2, 49–52, **58**, **60**, 61, 75, 81, 93, **161**, 238, 289n.4.3

equilibrium, 68, 69, 142, **163**

erosion, x, 65, 115, 181, 194, 195, **208**, 214, 224, 274

*Escherichia coli*, 102, 103, 192, 193, 206

Estonia, **247**

estuary, ix, 4, 10, 13, 16, 37, 106, 169, 176, 241–57

ethylenediaminetetraacetic acid, 138, 139

17α-ethynylestradiol, 113

*Eubosmina coregoni*, 26

Eurasian watermilfoil (*Myriophyllum spicatum*), 26

Europe, ix, x, 2, 5, 10, 14, 35, 39, 44, 111, 167, 176, 182, 185, 192, 199, 215, 236, 242, 249, 270, 288n.2.4, 290n.6.4, 291n.7.8

European ragweed (*Ambrosia*), 216

eutrophic lakes, 2, 3, 4, 5, 6, 7, 10, 12, 16, 20, 22, 31, 33, 35, 48, 76, 81, 82, 84, 158, **163**, 213, 224, 228, 230, 234, 235, 236, 238, 239, 240, 244, 265, 268

eutrophication, ix, x, 1, 5–13, 17, 21, 25, 27, 33, 37, 41, 42, 68, 73, 75, 82, 98, 99, 100, 102–110, 112, 115, 117, 118, 121, 123, **124**, 126, 128–132, 143, 155–177, 179–212, 217, 220, 224, 228, 229, 233–240, 241–257, 259–280, 288n.4.2, 289n.4.2

evaporation, 63, 182, 184, 185

evapotranspiration, 185

excrement, 37, 99, 180, 181, 197, 223, 269, 280

Experimental Lakes Area, x, **83**, 125, 155-58, 161, **163**, 170-72, 174, 180, 182-84, 199, 279, 288n.4.2, 289n.4.2

external loading, 234, 235, 238, 239, 254

Far East, 189, 199

farmland, *see* cropland

fats, 92, 97

fathead minnow (*Pimephales promelas*), 113, 173, 280

feedlots, 15, 115, 195, 267

fertilizer, 9, 10, 14, 15, 26, 27, 35, 37, 41, 42, 93-95, 99, 100, 103, 109, 111, 112, 115, 151, 158-60, **162**, 166, **168**, 169, 171, 174-76, 180, 181, 189, **190**, 194, 195, 197, 199, 206, 228, 235, 237, 242, 262, 263, 269, 271, 274, 278

fetch (lake measurement), **58**

Field Lake, 221, 223

filamentous algae, 3, 20, 32, 72, 168, 245, 247

fingerlings, 27, 84, 231

fingernail clams, 26

Finland, 236, **247**

First Nations, 32, 219, 266, 274, 277

fish, x, 2-5, 10, 12, 14, 16, 19, 20, 26, 27, 29, 30, 32, 35, 64, 78, 81, 83-85, 94, 96-98, 104, 106, 111, 114, 142, 153, 172, 174, 176, 195, 196, 199-204, 206, 211, 217, 220, 234, 236, 238-40, 263, 269, 272, 273, 277, 280, 291n.8.2

fish kill, 4, 24, 25, 30, 96, 237

fisheries, xi, 26, 27, 29, 30, 32, 33, 96, 98, 199, 203, 206, 211, 212, 219-21, 230, 231, 264, 268, 274, 275, 277

Fisheries Act, 269

Fisheries and Oceans (Canada), 200, 212, 264, 289n.4.2

Fisheries Research Board of Canada, 145, 155, 289n.4.2

Florida, 277

flushing time, 16, 17, 114

Food and Drug Administration (U.S.), 147

food chain, 1, 7, 77, 78, 80, 100, 118, 199, 200, 211, 245, 263

food webs, 6, 118, 156

forests, 8, 9, 22, 28, **36**, 42, 43, 70, 81, 113, 114, 160, 180, 184, 185, 187, 189, 200, **219**, 224, 228, 229, 237, 249, 250, 260

forest fire, x, 9, **183**

fossil fuels, 63, 67, 106, 151, 274

*Fragilaria*, 224

*Fragilaria capucina*, **228**

*Fragilaria crotonensis*, 224, **228**

France, 86, 106, 155, 239, 289n.5.1

freshwater, x, xi, 11, 12, 15, 41, 45, 77, 103, 111, 112, 177, 179, 180, 185, 188, 192, 195, 200-02, 206, 211, 212, 241, 244-47, 256, 264-67, 277, 280, 288n.4.2

freshwater drum (*Aplodinotus grunniens*), 26

Freshwater Institute (Canada), 143, 155, 159, 289n.4.2

frustules, 7, 20, 189, 216, 217, 224

fucoxanthin, 224, **229**

*Fucus*, 247

fungi, 16, 80, 81

full cost accounting, 211, 271

*Gammarus*, 84

Georgia Basin, 242

Germany, 106, 114, **247**

*Giardia*, 11

*Giardia lamblia*, 193

glacier, 8, 9, 45, 46, 182, 186, 187, 215

global positioning system, 203, 204

gold, 69, 87

golf courses, 180, 197, 223, 237, 267

GPS, *see* global positioning system

Grand Beach, **11**

Grand Drainage Canal, 108

grassland, 8, 9, 42, 70, 189, 190

Great Bear Lake, 2, 59, 81

Great Lakes, 10, 28, 30, 33, 119,
    120, 121, 157, 167, 180, **181**, 189,
    287n.2.2

Great Slave Lake, 2, 218

greenhouse gas, x, 106, 135, 137, 152

grey water, 279, 280

groundwater, 24, 45, 99, 107, 115, 142,
    187, 194, 196, 197, 198, 212, 229,
    268, 273, 288n.3.4

Gulf of Bothnia, **247**

Gulf of Mexico, 106, 241, 243

halocline, 245, 250

hand-auger, 203

hard water, 24, 69, 71–75, 128, 134

Hawaii, 260

Hayes River, 33

hayland, *see* grassland

Health Canada, 265

heavy metals, 142, 145, 149

hemoglobin, 145

hemolytic uremic syndrome, 208

hepatotoxin, 210

herbicides, 15, 101, 180, 197, 206, **208**,
    274

herbivores, 16, 77–80, **173**, 200–03,
    236

herring (*Clupea harengus*), 95, 96

*Hesperodiaptomus*, **79**

*Hexagenia*, 29, 30

Himmerfjärden, **247**

hogs, 3, 15, 112, 190, 192, 193, 195,
    **208**, 264

hormones, 102, 113, 156, 192, **208**,
    279, 280

Hot Lake, 59

House of Representatives (U.S.), 121,
    290n.7.2, 291n.7.7

Hudson Bay, 33, 218, 243

Hungary, 4

hydrogen, 38, 45, 46, 71, 72, **74**, 88, 90

hydrologic cycle, 44

hydroxyl ions, 71, 72, **74**, 75

hypereutrophic lakes, 4, 5, **57**

hypolimnion, 2–4, 49–52, 60, 61, 78,
    81, 82, 162, 183, 234, 238, 239, 240,
    289n.4.3

igneous rock, 8, 94

*ignis fatuus*, 91, 289n.5.2

IJC, *see* International Joint Commission

Illinois, **23**

Indian and Northern Affairs Canada,
    265

industry, ix, 11, 25, 28, 31, 38, 73, 84,
    95–98, 100, 103, 110, 112, 114, 129,
    131, 138, 139, 141, 146, 148, 150,
    152, 153, 157, 166, 185, 188, **190**,
    200, 211, 212, 223, 242, 248, 260,
    264, 267–69, 271, 278

infrared radiation, 52, 54, **57**, 58, 77,
    288n.3.2

Inland Waters Directorate (Canada),
    134

inorganic matter, **226**

Iowa, **23**

insects, 6, 53, 64, 76, 81, 95, 217

insecticides, 31, 95

internal loading, 26, 239, 256

plankton, 3, 20, 56, 76, 82, 117, 158, **228**, 240, 245

plants, 1–3, 5, 10, 13–16, 26, 39, 55, 63, 64, 68, 72, 75–81, 83, 84, 88, 89, 93, 99, 100, 106, 108–12, 114, 116–18, 129, 131, 139, 162, 187, 194, 196, **207**, 216, 235, 236, 239, 240, 261, 268, 269

plastics, 139, 151

PLUARG, *see* Pollution from Land Use Activities Research Group

point sources (of nutrients), 10, 15, 100, 180, 183, 184, 249, 251, 253, 263, 266, 267

pollen, 9, 20, 170

pollution, 119, 120, 123, 124, 131, 151, 187, 197, 212, 237, 246, 259–61, 270, 272, 275, 278, 288n.4.2, 290n.6.4, 291n.7.6

Pollution from Land Use Activities Research Group, 181

polycarboxylates, 137, 138, 141

polychlorinated biphenyls, 31

polymictic lakes, 240

*Pontoporeia*, 52

Portage la Biche, 218

potassium, 93, 126

potassium aluminum sulfate, 105, 111, 234, 239

poultry, 114

power plants, 131, 215, 275

Precambrian Shield, 216

precipitation, 38, 43, 45, 69, 94, 158, 163, 170, 180, 182–87, 189, 194, 208, 210, 263, 279

precipitates, 69, 70, 72–75, 83, 92, 93, 100, 101, 104, 105, 109, 111, 112, 127, 132, 138, 139, 180, 234, 289n.4.4

predators, 16, 26, 76, 77, 84, 173, 176, 200–02, 204–07, 211, 217, 238, 245

proteins, 42, 90, 92

protozoans, 11, 39, 70, 80

Puget Sound, 116

pyrophosphates, 127

Qu'Appelle Lakes, 231

radionuclides, x, 156, 162–65, 214, 215, 216, 223

radium, 164, 165, 214

radon, 162–65, 214

rain, *see* precipitation

rainbow smelt (*Osmerus mordax*), 29

rainbow trout (*Oncorhynchus mykiss*), 84, 85

Rainy River, 33

recycling, 21, 26, 76, 94, 101, 104, 108, 113, 114, 166, 245, 253, 269

Red Deer Brook, 218, 223

Red Deer River, 265

Red River, 32, 33, 35

Redfield ratio, 67

red herring, 95–97

relative thermal resistance, **51**

reservoir, x, 10, 16, 37, 113, 156, 183, 274

respiration, 64, 66, 67, 70, 75, 76, 81, 85, 92, 159, 161, 288n.4.1

Reuss Committee (U.S.), 121, 122, 127, 128, 129, 130, 132, 143

reverse osmosis, 40

Rhine River, 21

Rhode Island, 245

riparian zone, 41, 187, 190, 194–97, **208**, 263, 268, 273

River Norrström, 250, **252**

rivers, 16, 17, 44, 69, 99, 102, 114, 115, 180, 182, 187, 190, 194–98, 206, 212, 241, 243, 249, 264, 272

ribonucleic acid (RNA), 90, 91, 92

RNA, *see* ribonucleic acid

roach (*Rutilus rutilus*), 238

FIGURE 1.3: *Cyanobacterial scums washing ashore on the Grand Beach, Lake Winnipeg, August 2006.* Photograph by Lori Volkart.

FIGURE 2.2: *Lake Erie from the air in 1969, showing a massive bloom of Cyanobacteria or blue-green algae.* Photograph by DWS.

FIGURE 2.4: *A satellite photograph of Lake Winnipeg (right) in September 2005. Colours ranging from browns through yellows to greens indicate increasing vegetation density or vigour. In the case of Lake Winnipeg, greens are due to floating blue-green algae (i.e., surface blooms) almost indistinguishable from the green of the vegetation in the forest on either side of the lake. White indicates clouds. The surface blooms have reoccured for the last several years. Lakes Winnipegosis and Manitoba are to the left of the lake. The red-brown areas in the south basin of lakes Winnipeg and Manitoba are caused by silt suspended by wind, the result of its shallow depth. Similar observations have been made in all recent summers.*

Normalized difference vegetation index map supplied by Greg McCullough, University of Manitoba. Created from National Aeronautics and Space Administration MODIS (Moderate Resolution Imaging Spectroradiometer) bands 1 and 2 data.

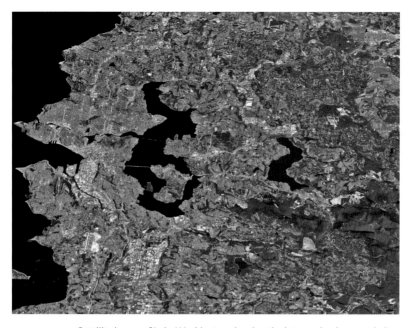

FIGURE 6.1: *Satellite image of Lake Washington showing the intense development in its catchment by the city of Seattle.*

Image from NASA Landsat Program, 2002, Landsat TM scene LT5046027000220210, Orthorectified and Terrain Corrected. USGS, Sioux Falls, SD, 07/21/2002.

FIGURE 9.1: *Left: Lake 227 before fertilization, with apparatus for several smaller experiments visible on the lake's surface. Right: Lake 227 in midsummer after fertilization began. Oligotrophic Lake 305 is shown in both pictures.*

FIGURE 9.5: *Lake 226 in midsummer 1973. The far basin (L226N) is being fertilized with phosphorus, nitrogen, and carbon. The near basin (L226S) is receiving only nitrogen and carbon and is in a near-pristine condition. The yellow line at the centre of the lake is a waterproof curtain separating the basins.* From Schindler (1974).

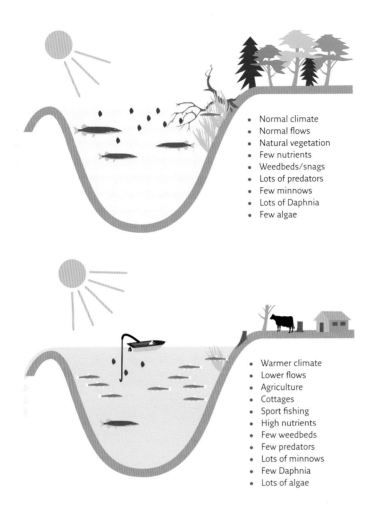

- Normal climate
- Normal flows
- Natural vegetation
- Few nutrients
- Weedbeds/snags
- Lots of predators
- Few minnows
- Lots of Daphnia
- Few algae

- Warmer climate
- Lower flows
- Agriculture
- Cottages
- Sport fishing
- High nutrients
- Few weedbeds
- Few predators
- Lots of minnows
- Few Daphnia
- Lots of algae

FIGURE 10.8: *A depiction of the cumulative effects of nutrient loading, land-use change, climate, and overfishing on the eutrophication of lakes. Top: A natural lake, with abundant top predators, normal water residence time, a catchment protected by natural vegetation, and normal climate. Note the clear blue water, the result of abundant zooplankton (represented here by* Daphnia, *the small red symbols). Bottom: The same lake after human habitation, land-use change, and agriculture have increased nutrient loads; overfishing has removed predators; fish habitat has been destroyed; and climate warming has reduced the outflow, causing increased retention of nutrients. Note that the water has become green, as a result of the combined effects, as described above.*

Drawing by Brian Parker and Lara Minja.

## Well Planned Watershed

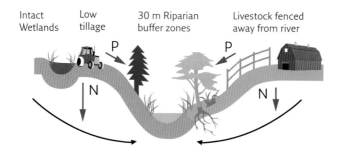

Intact Wetlands

Low tillage

30 m Riparian buffer zones

Livestock fenced away from river

## Poorly Planned Watershed

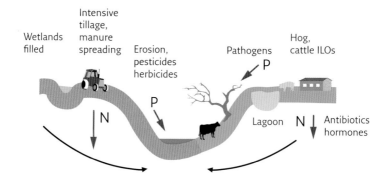

Wetlands filled

Intensive tillage, manure spreading

Erosion, pesticides herbicides

Pathogens

Hog, cattle ILOs

Lagoon

Antibiotics hormones

FIGURE 10.9: *Top: A well-planned watershed, where wetlands and riparian zones are protected, and various measures are used to keep nutrients from reaching lakes and streams. Bottom: The traditional way of treating a watershed, allowing development and agriculture to develop without any thought to water protection, is still generally used, because the science to support change is ignored by those who make land-use decisions.*
Drawing by Brian Parker and Lara Minja.

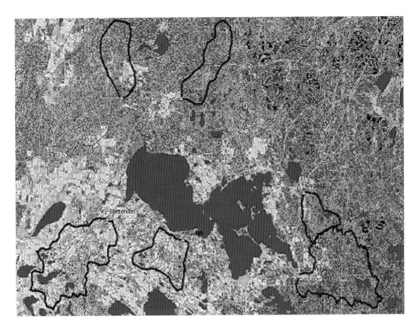

FIGURE 11.2: *Lac la Biche and its catchment. The red and green areas are forested, dominated by conifers and aspen, respectively. The white and gray areas south of the lake are cleared for agriculture. The town is shown as a gray blob in the southernmost tip of the lake. The black lines outline catchments studied for nutrient yield. The main inflow, the Owl River, enters from the northeast, draining largely untouched forests and small lakes. Of the portion of the basin shown, about 50% has been converted to agriculture or urban area.*
Figure from Neufeld (2005).

FIGURE 11.4: *The surface water of Lac la Biche near the mission, September 2005. The "green paint" colour is the result of a massive bloom of nitrogen-fixing Aphanizomenon.*
Photograph by DWS.

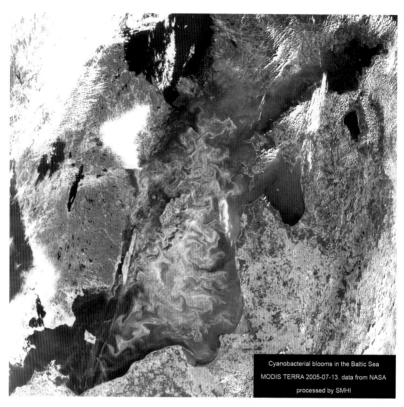

Cyanobacterial blooms in the Baltic Sea

MODIS TERRA 2005-07-13, data from NASA

processed by SMHI

FIGURE 13.3: *Cyanobacterial bloom in the Baltic Sea, July 13, 2005. The image is from the satellite sensor MODIS (Moderate Resolution Imaging Spectroradiometer).*

Satellite data is from NASA, GES Distributed Active Archive Center, and the data are processed by the SMHI.